Advances in Intelligent Systems and Computing

Volume 1149

The series "Advances in Intelligent Systems and Computing" contains publications on theory, applications, and design methods of Intelligent Systems and Intelligent Computing. Virtually all disciplines such as engineering, natural sciences, computer and information science, ICT, economics, business, e-commerce, environment, healthcare, life science are covered. The list of topics spans all the areas of modern intelligent systems and computing such as: computational intelligence, soft computing including neural networks, fuzzy systems, evolutionary computing and the fusion of these paradigms, social intelligence, ambient intelligence, computational neuroscience, artificial life, virtual worlds and society, cognitive science and systems, Perception and Vision, DNA and immune based systems, self-organizing and adaptive systems, e-Learning and teaching, human-centered and human-centric computing, recommender systems, intelligent control, robotics and mechatronics including human-machine teaming, knowledge-based paradigms, learning paradigms, machine ethics, intelligent data analysis, knowledge management, intelligent agents, intelligent decision making and support, intelligent network security, trust management, interactive entertainment, Web intelligence and multimedia.

The publications within "Advances in Intelligent Systems and Computing" are primarily proceedings of important conferences, symposia and congresses. They cover significant recent developments in the field, both of a foundational and applicable character. An important characteristic feature of the series is the short publication time and world-wide distribution. This permits a rapid and broad dissemination of research results.

**** Indexing: The books of this series are submitted to ISI Proceedings, EI-Compendex, DBLP, SCOPUS, Google Scholar and Springerlink ****

More information about this series at http://www.springer.com/series/11156

Phayung Meesad · Sunantha Sodsee
Editors

Recent Advances in Information and Communication Technology 2020

Proceedings of the 16th International Conference on Computing and Information Technology (IC^{2}IT 2020)

Springer

Editors
Phayung Meesad
Faculty of Information Technology and Digital Innovation
King Mongkut's University of Technology North Bangkok
Bangkok, Thailand

Sunantha Sodsee
Faculty of Information Technology and Digital Innovation
King Mongkut's University of Technology North Bangkok
Bangkok, Thailand

ISSN 2194-5357 ISSN 2194-5365 (electronic)
Advances in Intelligent Systems and Computing
ISBN 978-3-030-44043-5 ISBN 978-3-030-44044-2 (eBook)
https://doi.org/10.1007/978-3-030-44044-2

This Springer imprint is published by the registered company Springer Nature Switzerland AG
The registered company address is: Gewerbestrasse 11, 6330 Cham, Switzerland

Preface

In the digital era, computing and information technology play important roles to connect everything to support human beings. More and more new digital techniques have been continuously contributed to society. To keep up with the trends of new technology, the International Conference on Computing and Information Technology (IC^2IT) has been organized for 16 years. It serves the IT community as a platform for IT and computer science researchers to share new research and findings.

The main contributions of this volume are in the areas of data mining, data science, Internet of things, and smart information systems. The research contributions include big data, artificial intelligence, machine learning, natural language processing, speech recognition, image and video processing, as well as deep learning. These lead to the major research and engineering directions for autonomous driving, language assistant, automatic translation, and answering systems. Those changes also reflect economic changes in the world, which are increasingly dominated by the needs of enhanced globalization, worldwide cooperation (including its competitive aspects), and by emerging global problems.

The International Conference on Computing and Information Technology (IC^2IT) celebrates its 16th anniversary in 2020. To ensure high quality, 55 submissions from 12 countries were thoroughly reviewed by at least two, usually even three members of the international program committee (IPC). Through a positive vote of at least two IPC members, 20 submissions have been accepted for presentation at the conference and inclusion in the conference proceedings, which are published in the well-established and worldwide-distributed series on Advances in Intelligent Systems and Computing edited by Janusz Kacprzyk for eight years now.

Naturally, to unite these different experiences and approaches in one conference and one book requires a great preparation effort. Therefore, we would like to thank the authors as well as the members of the program committee, and all participating university collaborates in both Thailand and overseas for their outstanding support and academic cooperation. We also would like to say thank you to all staff members of the Faculty of Information Technology and Digital Innovation at

King Mongkut's University of Technology North Bangkok, who have carried out many technical and organizational tasks.

Finally, we are convinced that all authors, presenters, and participants will make IC^{2}IT 2020 a successful event. Therefore, we wish all of them a pleasant time in Pattaya and hope that the conference will find the right way for fruitful discussions of competitive ideas, in which sustainable new systems may incorporate.

February 2020

Phayung Meesad
Sunantha Sodsee

Organization

Program Committee

M. Aiello	UNI-Stuttgart, Germany
S. Auwatanamongkol	NIDA, Thailand
T. Bernard	Li-Parad, France
S. Boonkrong	SUT, Thailand
P. Boonyopakorn	KMUTNB, Thailand
A. Bunteong	UBU, Thailand
K. Chochiang	PSU, Thailand
N. H. H. Cuong	Danang Uni, Vietnam
T. Eggendorfer	HS Weingarten, Germany
M. Hagan	Okstate U, USA
S. Hiranpongsin	UBU, Thailand
W. Janratchakool	RMUTT, Thailand
N. Jongsawat	RMUTT, Thailand
M. Kaenampornpan	MSU, Thailand
M. Ketcham	KMUTNB, Thailand
P. Kropf	Uni NE, Switzerland
P. Kucharoen	NIDA, Thailand
P. Kunakornvong	RMUTT, Thailand
U. Lechner	UniBW, Germany
M. Maliyaem	KMUTNB, Thailand
C. Namman	UBU, Thailand
A. Nanthaamornphong	PSU, Thailand
K. Nimkerdphol	RMUTT, Thailand
W. Pacharoen	UBU, Thailand
P. Panakarn	UBU, Thailand
S. Pattanavichai	RMUTT, Thailand
W. Pongpech	NIDA, Thailand
N. Porrawatpreyakorn	KMUTNB, Thailand

P. Saengsiri	TISTR, Thailand
T. Siriborvornratanakul	NIDA, Thailand
S. Smanchat	KMUTNB, Thailand
W. Sriurai	UBU, Thailand
S. Tachaphetpiboon	PBRU, Thailand
C. Thaenchaikun	PSU, Thailand
J. Thongkam	MSU, Thailand
N. Tongtep	PSU, Thailand
K. Treeprapin	UBU, Thailand
D. Tutsch	Uni-Wuppertal, Germany
N. Utakrit	KMUTNB, Thailand
S. Valuvanathorn	UBU, Thailand
N. Wisitpongphan	KMUTNB, Thailand
K. Woraratpanya	KMITL, Thailand
P. Wuttidittachotti	KMUTNB, Thailand

Organizing Partners

In Cooperation with

King Mongkut's University of Technology North Bangkok (KMUTNB)
FernUniversitaet in Hagen, Germany (FernUni)
Chemnitz University, Germany (CUT)
Oklahoma State University, USA (OSU)
Edith Cowan University, Western Australia (ECU)
Hanoi National University of Education, Vietnam (HNUE)
Gesellschaft für Informatik (GI)
Mahasarakham University (MSU)
Ubon Ratchathani University (UBU)
Kanchanaburi Rajabhat University (KRU)
Nakhon Pathom Rajabhat University (NPRU)
Phetchaburi Rajabhat University (PBRU)
Rajamangala University of Technology Krungthep (RMUTK)
Rajamangala University of Technology Thanyaburi (RMUTT)
Prince of Songkla University, Phuket Campus (PSU)
National Institute of Development Administration (NIDA)
Council of IT Deans of Thailand (CITT)
IEEE CIS Thailand

Contents

An Empirical Study Towards the Intention to Use QR Code Payment in Champasak Province, Lao People's Democratic Republic

Charnsak Srisawatsakul(✉) and Waransanang Boontarig

Faculty of Computer Science, Ubon Ratchathani Rajabhat University, Ubon Ratchathani, Thailand
Charnsak@researcher.in.th, Waransanang.b@ubru.ac.th

Abstract. The QR code payment is a type of mobile payment that utilizes the use of a 2D barcode. It is transforming the economy of several countries into a cashless society. The Champasak Province, Lao People's Democratic Republic also implemented this service countrywide. However, the adoption rate of this technology is minimal. That is to say, consumers still preferred to use cash for their purchases rather than QR code payment. Therefore, this paper aims to find the factors affecting the intention to use the QR code payment system of the consumers in Champasak Province. The research model was proposed based on TAM. Three more extended variables were added to the TAM model including Perceived Usefulness, Hedonic Motivation, and Social Influence. This study collected data in the form of a paper-based survey. There were 411 datasets utilized to conduct the hypotheses testing using multiple linear regression. The results suggested that Perceived Usefulness and Social Influence are the two strongest significantly affect the intention to use the QR code payment system. The finding of this study should help the QR code payment system provider to design the system suitable for consumers in Champasak Province.

Keywords: QR code · Payment system · TAM · Laos · Mobile payment

1 Introduction

The number of smartphone users has remarkable growth in the last decade. The smartphones are now very affordable than ever before. Moreover, it equipped with high-performance processing speed, larger storage capacity, better camera, a lot of sensors, etc. Therefore, the smartphone becomes an important part of people's life. One of the most rapid growth services in the smartphone is mobile payment (m-payment). Mobile payment can be defined as "any wireless means to initiate, activate or confirm a payment" [1]. There are the two largest types of mobile payments in the market. Firstly, the Near Field Communications (NFC) technology requires the smartphone equipped with NFC sensors. For example, the services from Apple Pay and Samsung Pay Secondly, the Quick-Response (QR) code payment system. The QR Code requires a smartphone that connected to the internet with an autofocus camera and mobile applications such as Alipay, WeChat Pay, etc. This paper focuses mainly on using the QR Code technology as a method for a mobile payment system. In this case, the

P. Meesad and S. Sodsee (Eds.): IC²IT 2020, AISC 1149, pp. 1–10, 2020.
https://doi.org/10.1007/978-3-030-44044-2_1

payment can simply be done with a simple scan of the QR Code using the provider's mobile application.

The QR payment system is transforming numerous countries to the cashless society. China is targeting to become the world's first cashless society by adopting the QR Code Payment by 2023 [2]. The most popular mobile payment services in China are Alipay and WeChat Pay, accounting for 87% and 76% of the market, accordingly [3]. On the other hand, the NFC payment systems have less popularity, the Apple Pay was used only 18% and 1% for Samsung Pay.

The Lao People's Democratic Republic is a developing country in Southeast Asia. The use of QR code payment services in Laos still not widely accepted by the end-users. Moreover, there is still a research gap in the adoption of the QR payment system in Laos. Therefore, the main purpose of this study is to explain factors affecting the intention to use the QR payment system of the consumer in Laos. To accomplish the desired objective, we have proposed a research model that is a modified version of the Technology Acceptance Model (TAM) with 3 factors related to QR code payment. This study has used TAM because it is a well-recognized model [4] to explain information technology adoption. Furthermore, TAM provides a general framework that allows the researcher to modify and test with other factors that influence the adoption in a different context.

The next section provides a review of related literature. Section 3 presents the hypotheses and research model used in this study. Section 4 provides the results from our hypotheses testing with statistical software. Section 5 provides a conclusion, contribution, implication of this study.

2 Related Literature

2.1 QR Code Payment System

The beginning of QR code started from the barcode which stores information on the horizontal dimension. An original barcode was able to record up to 13 characters. The QR Code is one type of 2-dimensional barcode developed by Masahiro Hara under the company name Denso Wave [5, 6]. The QR code store the data in both horizontal and vertical dimension. It could store data up to 7,000 characters which make them usable in countless context. Nowadays, Smartphones can be used to read QR code without the need of any QR code reader anytime and anywhere. However, the implementation of the QR code that has the biggest impact is the QR code payment system [7]. Consumers can easily transfer money to other individuals by scan the QR code within mobile applications. It helps to reduce the burden of cash maintenance. China is the country where the QR code payment is widely adopted. Most of the people in China prefer to transfer money with QR codes payment rather than cash [8]. Importantly, QR code payment is more acceptable than Apple Pay or Samsung Pay which uses NFC technologies [3]. On the opposite, the adoption of the QR code payment system in Laos is very low although the service has been offered for the consumer there for a short while [9], yet it is relatively not used, thus opening up the opportunity to work in this particular area.

2.2 Technology Acceptance Model

In the past decades, researchers have attempted to explain user behavior related to the adoption of technologies using research models. One of the oldest models is the Theory of Reasoned Action (TRA) [10]. The TRA explains that attitude and subjective norms will have an effect on behavioral intention. It is later updated with the Theory of Planned Behavior (TPB) [11] by adding the perceived behavioral control factor. The TRA and TPB were used as general behavioral intention models which are not particularly useful in the information technology context. Technology Acceptance Model (TAM) [4] is a model based on TRA that designed to explain behavioral intention in the information technology context. It is one of the most widely applied among researchers in the field of information technology acceptance study.

3 Research Methodology

3.1 Research Model

The research model used in this study is shown in Fig. 1 below.

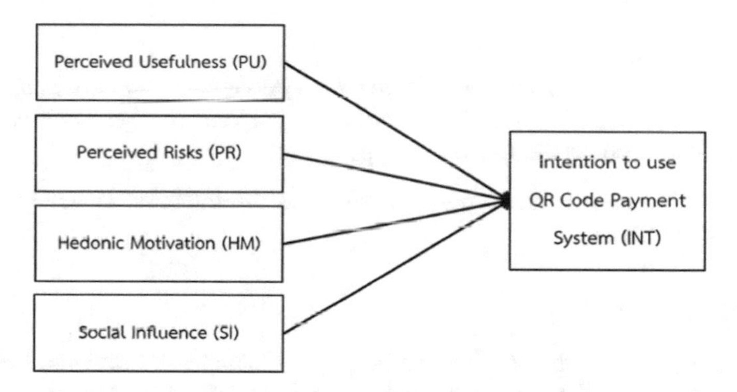

Fig. 1. The research model in this study.

The research model extends 3 more variables and removes perceived ease of use factor within the TAM factors for predicting the intention to use QR code payment in Laos. The reason that we remove perceived ease of use is that the QR code payment application is a simple type of application that is already easy enough for anyone who can use the smartphone to operate. The intention to use the QR code payment system is a dependent variable. Perceived Usefulness (PU) [4], Perceived Risks (PR) [12], Hedonic Motivation (HM) [13] and Social Influence (SI) [10] are independent variables used to predict the dependent variable The explanation and hypothesis of each variable are shown below.

Perceived Usefulness (PI)

PU was originally proposed in TAM. It can be defined as *"the degree to which a person believes that using a particular system would enhance his/her job performance"* [4].

In this context, if the users find that the QR code payment is useful to them, the intention to use the QR code payment will increase. Thus, the following hypothesis:

H1: Perceived Usefulness (PU) has a positive effect on the intention to use the QR code payment system (INT).

Perceived Risks (PR)
Perceived Risks was proposed by Featherman and Pavlou [12]. They have classified the risks into 7 categories. Performance risk, Financial risk, Time risk, Psychological risk, Social risk, Privacy risk, and Overall risk. The QR code payment system mainly involves financial and privacy risks. Therefore, it can be hypothesized as:

H2: Perceived financial and privacy risk (PR) have a positive effect on the intention to use the QR code payment system (INT).

Hedonic Motivation (HM)
Hedonic motivation is defined as *"the fun or pleasure derived from using a technology"* [13]. Researchers found that hedonic play an important role in determining technology acceptance and use directly.

H3: Hedonic Motivation (HM) has a positive effect on the intention to use the QR code payment system (INT).

Social Influence (SI)
Social Influence can be explained as *"the person's perception that most people who are important to him think he should or should not perform the behavior in question"* These lead us to propose the following hypotheses:

H4: Social Influence (SI) has a positive effect on the intention to use the QR code payment system (INT).

3.2 Data Collection

The target population of this study is the people in Champasak Province, Lao People's Democratic Republic who use the smartphone in their daily life. However, the number of populations is unknown. This research, therefore, uses over 400 samples to describe the entire population, according to Yamane's theory [14]. The sample of this research was done by accidental sampling method.

Data was collected using a self-administered questionnaire in which each item is measured on a 5-point Likert scale. The constructs of each factor were developed from the literature review. The questionnaires were sent to different public places such as restaurants, beverage stores, universities, fresh market etc. The survey was conducted for a period of 2 weeks in order to ensure that the participants got a similar experience of the QR code payment system.

In the questionnaire, there was an explanation of the QR code payment system such as the pictures showing usage of QR codes to send or receive payments. It is divided into 2 parts; the first part is about demographic information. The second part is the construct question variables. The questionnaire was translated from Thai into Laos language by a Laos native speaker.

The collected data were recorded as a Microsoft Excel spreadsheet for data screening. The Statistical Package for Social Science (SPSS) program was used to analyze the effect between dependent and independent variables using multiple linear regression.

4 Results

Out of the 459 questionnaires initially collected, 48 cases were rejected due to incomplete responses, thus giving a final participant of the 411 dataset that were you for empirical analysis.

4.1 Validity and Reliability Analysis

Before doing the actual data analysis, we had verified that the questionnaire had enough validity and reliability. For the validity analysis, the questionnaire was analyzed by using the Index of Item Objectives Congruence (IOC) analysis through 3 experts. Each question has an IOC value of more than 0.5, Hence, the questions considered as valid. Cronbach's alpha coefficient [15] was used to test the reliability of the questionnaire. The value of the alpha coefficient of the questionnaire which greater than 0.6 will consider that questionnaire as reliable. Cronbach's alpha coefficient of this study is 0.747, which considered being a good level of reliability. Therefore, the validity and reliability of this questionnaire are satisfied.

4.2 Demographic Variables

In a total of 411 participants, Tables 1 and 2 show the details of the demographic profile of gender and age groups of the participants. The participants had a smaller number of males than females. The majority of the people belong to the age group of 21–28.

Table 1. Gender of participants.

Gender	Frequently	Percentage
Male	125	30.4
Female	286	69.6
Total	**411**	**100.0**

Table 2. Age groups of participants.

Age groups	Frequently	Percentage
21–28	294	71.5
29–36	79	19.2
37–44	32	7.8
More than 44	6	1.5
Total	**411**	**100.0**

4.3 Correlation Analysis

Correlation analysis is a method to identify the relationship between the 2 variables. In this study, Pearson's Correlation Coefficient (R) was used as an indicator of the relationship between variables. The value of Pearson's correlation coefficient is between −1.0 and +1.0. If the value is close to −1.0, it means that both variables are indirectly correlated. On the other hand, if the value is near +1.0, it means both variables are directly related. Moreover, if the value is 0, it means that both variables are not related to one another at all.

A Pearson correlation coefficient was computed to assess the relationship between each variable. The results of Pearson's correlation shown in Table 1. It is indicated that there was no high correlation between 5 variables ($r < 0.7$, $n = 411$, $p < 0.05$) Therefore, the data can be used for further linear regression analysis. The highest correlation coefficient is between PU and SI ($r = 0.605$, $n = 411$, $p < 0.05$) (Table 3).

Table 3. Correlation analysis between variables.

Variables	PU	PR	HM	SI	INT
PU	1	.204**	.397**	.605**	.516**
PR	.204**	1	.227**	.220**	.192**
HM	.397**	.227**	1	.547**	.324**
SI	.605**	.220**	.547**	1	.524**
INT	.516**	.192**	.324**	.524**	1

** $P < 0.05$

4.4 Multicollinearity Analysis

In order to conduct the multiple linear regression analysis, the variables must not have the collinearity with one another. The multicollinearity could be detected by the value of the Variance inflation factors (VIF) and Tolerance. Table 4 shows the results of VIF and Tolerance for each variable. The highest VIF value in all the variables is 1.922 and the lowest value of Tolerance is 0.520. The value of VIF must not greater than 5. The value of Tolerance must greater than 0.2. Hence, the data in the study show no multicollinearity and could conduct multiple linear regression analysis.

Table 4. The value of tolerance and variance inflation factor (VIF).

Variables	Tolerance	Variance inflation factor (VIF)
PU	.624	1.603
PR	.929	1.076
HM	.684	1.463
SI	.520	1.922

**Dependent Variable: INT

4.5 Results from Multiple Linear Regression

Assumption Testing

The data were tested with 3 assumptions testing methods: homoscedasticity, linearity, and normality. The results of those analyses shown in Figs. 2, 3 and 4. The histogram of standardized residuals (Fig. 2) suggested that the data contained approximately normally distributed. This result supported by the normal P-P plot of standardized residuals (Fig. 3), which showed the value go along with the diagonal line without any systematic departures or substantial. Hence, there is a linear relationship with standardized residuals by predicted values. Therefore, the testing of normality and linearity were satisfied.

The test of homoscedasticity present in the scatterplot of standardized predicted values (Fig. 4). It showed the values of regression standardized residual were in the range of −3.3 to 3.3 without any problem with the outliner. Therefore, the data met the assumptions of homoscedasticity of variance and linearity.

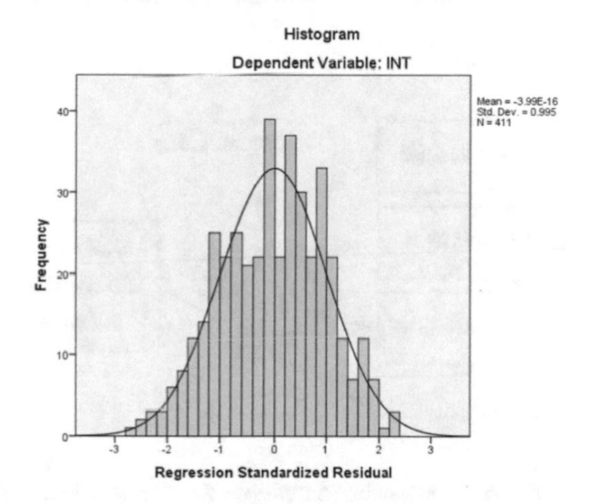

Fig. 2. Histogram of standardized residuals.

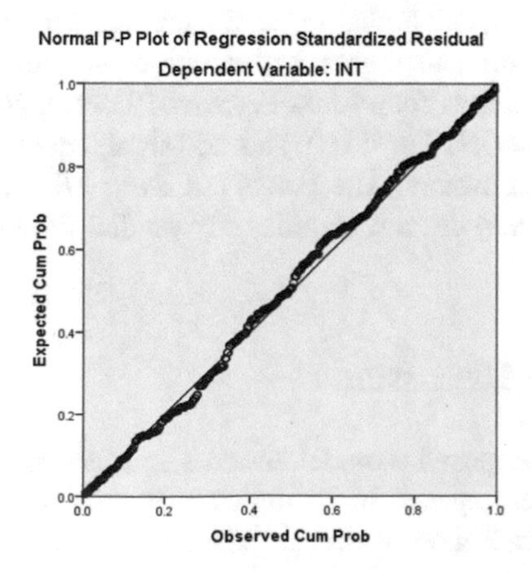

Fig. 3. Normal P-P plot of regression standardized residual.

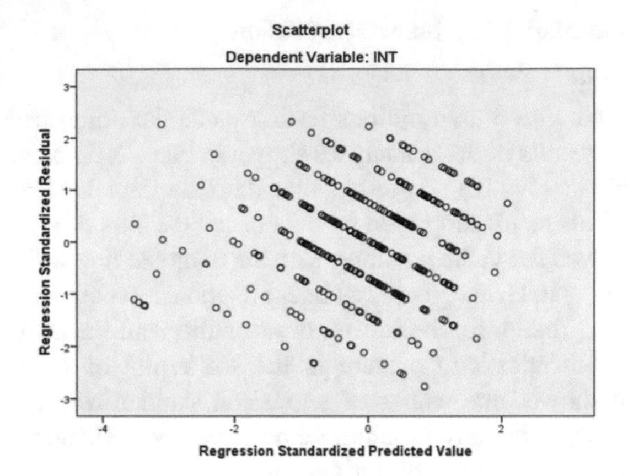

Fig. 4. Scatterplot of standardized predicted values.

Regression Variate Results

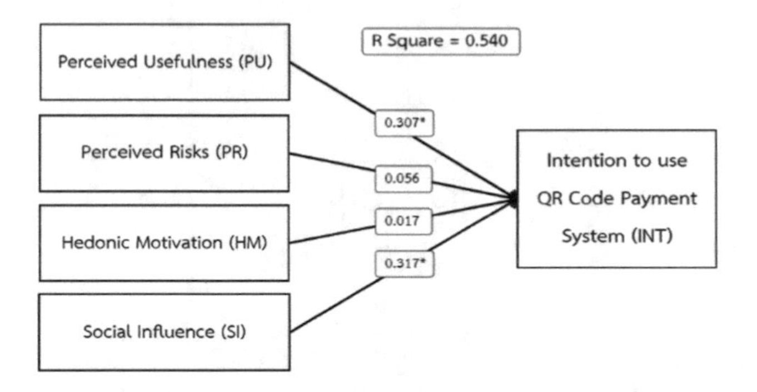

Fig. 5. The results of multiple linear regression.

Multiple linear regression was calculated to see if Perceived Usefulness, Perceived Risks, Hedonic Motivation, and Social Influence predicted the intention to use QR code payment systems. A significant regression equation was found with an R^2 of 0.54. Using the enter method it was found that Perceived Usefulness ($B = 0.307$, $p < 0.05$) and Social Influence ($B = 0.317$, $p < 0.05$) are the significant predictors of the intention to use QR code payment system. The Perceived Risks ($B = 0.056$, *n.s*) and Hedonic Motivation ($B = 0.017$, *n.s*) did not significantly predict the intention to use QR code payment systems.

5 Conclusion and Discussion

In this paper, we have proposed a model aims to find factors affecting the intention to use the QR code payment system in Champasak Province. The proposed model was based on TAM with extra 3 more variables. The summary of hypotheses testing results is shown in Table 5.

Table 5. Hypothesis testing results.

No	Hypotheses	Results
H1	PU → INT	Accept
H2	PR → INT	Reject
H3	HM → INT	Reject
H4	SI → INT	Accept

The results show that Perceived Usefulness and Social Influence are 2 factors that significantly affect the intention to use the QR code payment system in Champasak Province. This finding is consistent with the study of Liébana-Cabanillas et al. [16] Sang Ryu and Murdock [17], which suggested that Perceived Usefulness has the positive effect on the attitude to use QR Code. This could be interpreting that end users will adopt the QR code payment system only when they could perceive the usefulness in it. Therefore, the provider should be considered to deliver the usefulness of the system to consumers in marketing communication. It should help increase the adoption rate.

This unexpected finding suggests that the Perceived Risk and Hedonic Motivation were not significant at all. The participants may think the hedonic effect in the QR code payment is unnecessary because the mobile payment application is very simple. They also do not think that the risk of financial and personal privacy will affect their intention to use the QR code payment system. A possible explanation for this might be that they already trusted their financial institute because the QR payment systems in Laos run by the bank. This is an important issue for future research.

In this research, we concentrated on the research area in Champasak Province. In future work, we plan to investigate broadly into another province of Laos. Furthermore, the detailed comparison between the adoption of another type of mobile payments such as NFC is also interesting.

Acknowledgment. This work is partially supported by Ubon Ratchathani Rajabhat University. The authors would like to thank Ms. Boualaphan Souphanthong from Pakse Teacher Training College for her help in questionnaire translation and data collection.

References

1. Geva, B.: The wireless wire: do M-Payments and UNCITRAL model law on international credit transfers match raw? Bank. Financ. Law Rev. **27**, 249–264 (2014)
2. Morris, H.: China's march to be the world's first cashless society. https://www.straitstimescom/asia/east-asia/chinas-march-to-be-the-worlds-first-cashless-society-china-daily-contributor. Accessed 29 Dec 2019
3. Buchholz, K.: China's most popular digital payment services. https://www.statista.comchart/17409/most-popular-digital-payment-services-in-china/ Accessed 29 Dec 2019
4. Davis, F.D.: Perceived usefulness, perceived ease of use, and user acceptance of information technology. MIS Q. **13**(3), 319–340 (1989)
5. Soon, T.J.: QR code. Synth. J. **2008**, 59–78 (2008)

6. Denso Wave: History of QR Code. https://www.qrcode.com/en/history/. Accessed 29 Dec 2019
7. Lu, L.: Decoding Alipay: mobile payments, a cashless society and regulatory challenges. Butterworths J. Int. Bank. Financ. Law **33**(1), 40–43 (2017)
8. Minter, A.: China's Cashless Revolution. https://www.bloomberg.com/opinion/articles/2017-07-19/china-s-cashless-revolution. Accessed 29 Dec 2019
9. LaoVietBank: LaoVietBank launches QR Pay service on LaoVietBank Digital Banking application. https://www.laovietbank.com.la/files/img_slider/PRpay.png. Accessed 10 Apr 2019
10. Fishbein, M., Ajzen, I.: Belief, Attitude, Intention, and Behavior: An Introduction to Theory and Research. Addison-Wesley, Reading (1975)
11. Ajzen, I.: The theory of planned behavior. Organ. Behav. Hum. Decis. Process. **50**, 179–211 (1991)
12. Featherman, M.S., Pavlou, P.A.: Predicting e-services adoption: a perceived risk facets perspective. Int. J. Hum. Comput. Stud. **59**, 451–474 (2003)
13. Venkatesh, V.: Consumer acceptance and user of information technology: extending the unified theory of acceptance and use of technology. MIS Q. **36**, 157–178 (2012)
14. Yamane, T.: Statistics: An Introductory Analysis, 3rd edn. Harper & Row, New York (1973)
15. Santos, J.R.A.: Cronbach's alpha: a tool for assessing the reliability of scales. J. Ext. **37**, 1–5 (1999)
16. Liebana-Cabanillas, F., Ramos de Luna, I., Montoro-Rios, F.J.: User behaviour in QR mobile payment system: the QR payment acceptance model. Technol. Anal. Strateg. Manag. **27**, 1031–1049 (2015)
17. Ryu, J.S., Murdock, K.: Consumer acceptance of mobile marketing communications using the QR code. J. Direct Data Digit. Mark. Pract. **15**, 111–124 (2013)

Identifying of Decision Components in Thai Civil Case Decision by Text Classification Technique

Jantima Polpinij[1(✉)], Poramin Bheganan[2], Bancha Luaphol[3(✉)], Chumsak Sibunruang[1], and Khanista Namee[4]

[1] Intellect Laboratory, Faculty of Informatics, Mahasarakham University, Mahasarakham, Thailand
jantima.p@msu.ac.th, inspire.ch@gmail.com

[2] Mahidol University International College, Mahidol University, Nakhon Pathom, Thailand
poramin.bhe@mahidol.edu

[3] Department of Digital Technology, Kalasin University, Kalasin, Thailand
bancha.lu@ksu.ac.th

[4] King Mongkut's University of Technology North Bangkok, Prachinburi Campus, Prachinburi, Thailand
khanista.n@fitm.kmutnb.ac.th

Abstract. A Thai civil case decision document is typically presented in a semi-structured form. Generally, Thai civil case decision documents consist of four major components comprising the dispute, facts, decision, and judgment. To perform text summarization or information extraction on this document, the first process should recognize major components in the Thai civil case decision document. This has not been addressed previously and becomes the challenge for our study that aims to present a method of identifying the four major components utilizing the text classification technique. We employed two weighting schemes and three supervised machine learning algorithms and downloaded the dataset from the Supreme Court of Thailand website (http://www.supremecourt.or.th). After testing by recall, precision, and F1 satisfactory results were achieved for identifying major components in Thai civil case decision documents at 0.83, 0.80 and 0.81, respectively.

Keywords: Thai civil case decision document · Decision components · Text classification

1 Background

A lawsuit can be described as a means to resolve an argumentation between two parties [1]. There are two major types of lawsuits as legal actions involving criminal cases and civil cases. Civil cases usually relate to broken contracts or

P. Meesad and S. Sodsee (Eds.): IC^{2}IT 2020, AISC 1149, pp. 11–20, 2020.
https://doi.org/10.1007/978-3-030-44044-2_2

torts that are wrongful acts that result in damage or injury. Civil cases are usually brought to court by way of an action. Once a case is finalized in the Supreme Court, a document is produced detailing the manner of settlement. This document is called a court decision. These court decisions comprise data sources of knowledge for legal professionals as well as the general public.

In Thailand, each civil case involves a documented decision called a civil decision document. This document details the vital components of the court case including the dispute, facts, decision, and judgment. Generally, the fact component describes the situation, persons, places and surrounding evidence, while the dispute component describes the causes and reasons for the lawsuit. The decision component involves statements made by the judges analyzing each cause. The last component of the Thai civil case decision document is the judgment. This describes the final judgment given by the Appeals Court judge in the case. It is easier to understand and comprehend the legal aspects of Thai civil case decision documents if each decision component is correctly distinguished.

Unfortunately, Thai civil case decision documents are not written following a specific pattern and readers are often unable to clearly distinguish each decision component. This is the challenge for our study. Developing a tool to identify each component of Thai civil case decision documents may help in the understanding and recognition of specific aspects. Here, we present a method for the automatic identification of decision components in Thai civil case decision documents.

The rest of the paper is organized as follows. We discuss the related work in Sect. 2. Section 3 describes the dataset. Section 4 explains the method of classifier modeling. Section 5 presents the experimental results. Finally, Sect. 6 concludes this study.

2 Related Work

Handling Thai civil case decision documents using computer technology can be very useful in facilitating the search and analysis of legal cases involving civil law in Thailand. Here, we present a method for automatic identification of the components in Thai civil case decisions using a text classification technique. Many previous studies have addressed the use of text classification to categorize legal documents in other countries [2–15].

In 2009, Palau and Moens [2] proposed maximum entropy and the Naïve Bayes (NB) to classify the proposition of documents extracted from two different corpora. The first corpus contained legal texts of the European Court of Human Rights (ECHR) while the second, the Araucaria corpus, used only the structured set in English collected and analyzed according to a specific methodology as part of a project at the University of Dundee (UK). Each document was classified into argumentative or non-argumentative before further processing. In 2010, de Maat *et al.* [3] compared the Support Vector Machine (SVM) algorithm with a pattern-based approach to classify sentences in Dutch legislation into definition, obligation, repeal, application and provision. They reported that a pattern-based classifier was more robust for legal document categorization at the sentence level; however, their obtained classifiers generalized poorly on new legislation and cross-domain modeling.

Similar to Boella *et al.* [4], they also used the SVM algorithm to classify related legal domains that fitted specific legal text from Italian national law documents. They also applied Information Gain-based feature selection to improve the performance of the classifier. After extensive experimentation, they found that the similarity of topics in the legal documents and the short text extractions were problematic for accurate classification.

Farzindar and Lapalme [5] presented an approach to summarize legal proceedings in the federal courts of Canada. They displayed the legal proceeding summary in table format. They used the identification of a thematic structure for documents and the determination of argumentative themes as textual units in the judgment to extract important sentences. This generated a summary of four themes as introduction, context, juridical analysis, and conclusion.

Hachey and Grover [6] also proposed a classifier to determine the rhetorical status of sentences in texts from the House of Lord's Judgment (HOLJ) corpus. The extracted sentences were based on the feature set of Teufel and Moens [7]. They identified seven rhetorical roles as facts, proceedings, background, framing, disposal, textual, and others. Their experiments were conducted with four classifiers, i.e. C4.5, NB, SVM, and Winnow using the same features. The C4.5 returned better results in terms of micro-averaged F1 scores as 65.4 using location features only.

Meanwhile, Schilder and Molina-Salgado [8] presented the influence of repetitive legal phrases in the text using a graph-based approach. They used a similarity function between sentences to represent the legal text and assumed that some of the paragraphs could summarize the whole document or some parts of it. To identify the type of paragraphs, they computed inter-paragraph similarity scores and selected those paragraphs that best matched with other paragraphs. Their proposed system operated similarly to a voting system, wherein each paragraph assigns a vote for another paragraph. The paragraphs with the highest voting score were selected as the summary. Phrase similarity was used to calculate the voting score, and calculated by the co-occurrence of phrases between each pair of paragraphs. Longer matched phrases received a higher score.

More recently, Sulea *et al.* [14] and Howe *et al.* [15] also applied text classification to legal documents. They tried to discover, compare and use more than one classification technique to enable better performance. Their results indicated that some techniques could not be applied to the next stage of legal document summarization and information extraction. Before applying these techniques, it would be better if it was possible to recognize the major components in legal documents, especially the major components in civil case decision documents. As a result, this problem is addressed here by applying text classification to identify the major components in civil case decision documents.

However, although the mention studies give an idea that can work on legal document, those works may not be able to apply for the next stage of some area studies such as legal document summarization, information extraction for legal document, and so on. This is because before working on those studies, it would be better if we can recognize for major components in the legal document, especially the major components in the civil case decision document. As the result, this problem is addressed in this study by applying text classification to identify the major components in the civil case decision document study.

3 Dataset

The dataset used in this study downloaded from Supreme Court of Thailand Website (http://www.supremecourt.or.th). 2,000 civil case decision documents are retrieved on August - October 2019. Each civil case decision document are formatted as a text file with UTF-8 encoded in Thai language. It consists of four major components i.e. dispute, fact, decision, and judgment. An example of civil case decision document is shown as Fig. 1.

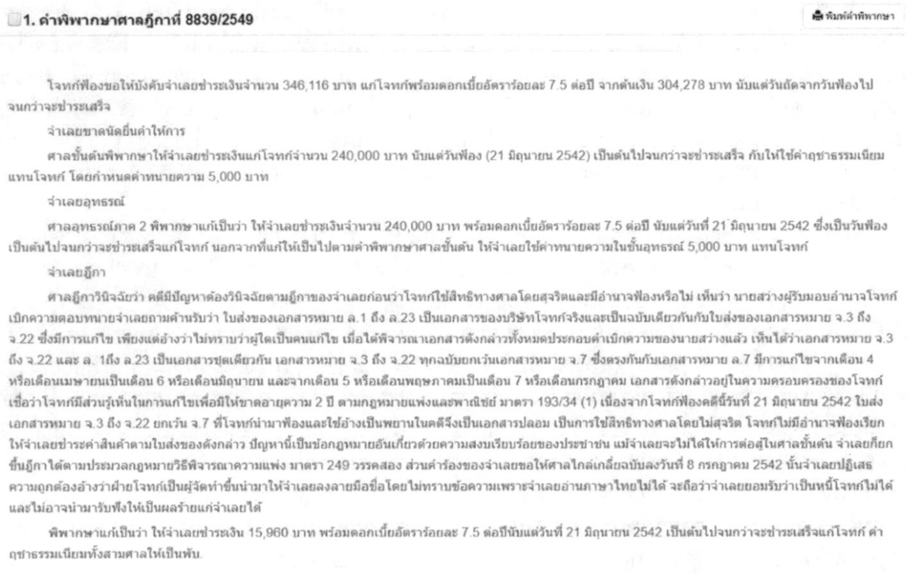

1. คำพิพากษาศาลฎีกาที่ 8839/2549 พิมพ์คำพิพากษา

โจทก์ฟ้องขอให้บังคับจำเลยชำระเงินจำนวน 346,116 บาท แก่โจทก์พร้อมดอกเบี้ยอัตราร้อยละ 7.5 ต่อปี จากต้นเงิน 304,278 บาท นับแต่วันถัดจากวันฟ้องไปจนกว่าจะชำระเสร็จ

จำเลยขาดนัดยื่นคำให้การ

ศาลชั้นต้นพิพากษาให้จำเลยชำระเงินแก่โจทก์จำนวน 240,000 บาท นับแต่วันฟ้อง (21 มิถุนายน 2542) เป็นต้นไปจนกว่าจะชำระเสร็จ กับให้ใช้ค่าฤชาธรรมเนียมแทนโจทก์ โดยกำหนดค่าทนายความ 5,000 บาท

จำเลยอุทธรณ์

ศาลอุทธรณ์ภาค 2 พิพากษาแก้เป็นว่า ให้จำเลยชำระเงินจำนวน 240,000 บาท พร้อมดอกเบี้ยอัตราร้อยละ 7.5 ต่อปี นับแต่วันที่ 21 มิถุนายน 2542 ซึ่งเป็นวันฟ้องเป็นต้นไปจนกว่าจะชำระเสร็จแก่โจทก์ นอกจากที่แก้ให้เป็นไปตามคำพิพากษาศาลชั้นต้น ให้จำเลยใช้ค่าทนายความในชั้นอุทธรณ์ 5,000 บาท แทนโจทก์

จำเลยฎีกา

ศาลฎีกาวินิจฉัยว่า คดีมีปัญหาต้องวินิจฉัยตามฎีกาของจำเลยก่อนว่าโจทก์ใช้สิทธิทางศาลโดยสุจริตและมีอำนาจฟ้องหรือไม่ เห็นว่า นายสว่างผู้รับมอบอำนาจโจทก์เบิกความตอบทนายจำเลยถามค้านรับว่า ใบส่งของเอกสารหมาย ล.1 ถึง ล.23 เป็นเอกสารของบริษัทโจทก์จริงและเป็นฉบับเดียวกันกับใบส่งของเอกสารหมาย จ.3 ถึง จ.22 ซึ่งมีการแก้ไข เพียงแต่อ้างว่าไม่ทราบว่าผู้ใดเป็นคนแก้ไข เมื่อได้พิจารณาเอกสารดังกล่าวทั้งหมดประกอบคำเบิกความของนายสว่างแล้ว เห็นได้ว่าเอกสารหมาย จ.3 ถึง จ.22 และ ล. 1ถึง ล.23 เป็นเอกสารชุดเดียวกัน เอกสารหมาย จ.3 ถึง จ.22 ทุกฉบับยกเว้นเอกสารหมาย จ.7 ซึ่งตรงกันกับเอกสารหมาย ล.7 มีการแก้ไขจากเดือน 4 หรือเดือนเมษายนเป็นเดือน 6 หรือเดือนมิถุนายน และจากเดือน 5 หรือเดือนพฤษภาคมเป็นเดือน 7 หรือเดือนกรกฎาคม เอกสารดังกล่าวอยู่ในความครอบครองของโจทก์ เชื่อว่าโจทก์มีส่วนรู้เห็นในการแก้ไขเพื่อมิให้ขาดอายุความ 2 ปี ตามกฎหมายแพ่งและพาณิชย์ มาตรา 193/34 (1) เนื่องจากโจทก์ฟ้องคดีนี้วันที่ 21 มิถุนายน 2542 ใบส่งเอกสารหมาย จ.3 ถึง จ.22 ยกเว้น จ.7 ที่โจทก์นำมาฟ้องและใช้อ้างเป็นพยานในคดีจึงเป็นเอกสารปลอม เป็นการใช้สิทธิทางศาลโดยไม่สุจริต โจทก์ไม่มีอำนาจฟ้องเรียกให้จำเลยชำระค่าสินค้าตามใบส่งของดังกล่าว ปัญหานี้เป็นข้อกฎหมายอันเกี่ยวด้วยความสงบเรียบร้อยของประชาชน แม้จำเลยจะไม่ได้ให้การต่อสู้ในศาลชั้นต้น จำเลยก็ยกขึ้นฎีกาได้ตามประมวลกฎหมายวิธีพิจารณาความแพ่ง มาตรา 249 วรรคสอง ส่วนคำร้องของจำเลยขอให้ศาลไกล่เกลี่ยฉบับลงวันที่ 8 กรกฎาคม 2542 นั้นจำเลยปฏิเสธความถูกต้องอ้างว่าฝ่ายโจทก์เป็นผู้จัดทำขึ้นนำมาให้จำเลยลงลายมือชื่อโดยไม่ทราบข้อความเพราะจำเลยอ่านภาษาไทยไม่ได้ จะถือว่าจำเลยยอมรับว่าเป็นหนี้โจทก์ไม่ได้ และไม่อาจนำมารับฟังให้เป็นผลร้ายแก่จำเลยได้

พิพากษาแก้เป็นว่า ให้จำเลยชำระเงิน 15,960 บาท พร้อมดอกเบี้ยอัตราร้อยละ 7.5 ต่อปีนับแต่วันที่ 21 มิถุนายน 2542 เป็นต้นไปจนกว่าจะชำระเสร็จแก่โจทก์ ค่าฤชาธรรมเนียมทั้งสามศาลให้เป็นพับ

Fig. 1. An example of civil case decision document

After having the dataset, this data is performed by some domain experts to manually identify the major components in the civil case decision document such as dispute, fact, decision, and judgment. Later, each civil case decision document is separated to become four documents. The documents with the same mentioned component will be grouped together. Finally, we obtain the datasets that will be used for our dataset for this study. Our dataset can be summarized in Table 1.

For this dataset, it is used in our experiment to obtain the classifier models based on 10-fold cross validation. However, we provide 100 civil case decision documents to evaluate the obtained classifier models for identifying the major components in each civil case decision document in the stage of model usage.

Table 1. Summary of the dataset

Component description	Total of document
Dispute	500
Fact	500
Decision	500
Judgment	500

4 The Proposed Method

The proposed method consists of two major stages: civil case decision document pre-processing and classifier modeling of each civil case decision component. Each stage can be detailed as follows.

4.1 Civil Case Decision Document Pre-processing

We start with the process of word segmentation. This process intends to separate text into a meaning unit called as "*words*". This study applies a dictionary-based word segmentation with longest matching algorithm to segment words as it is well-known that the accuracy of the mentioned algorithm is acceptable [16].

After segmenting words, the text is cleansed by removing special characters and punctuations such as "\n", "(" ,")", "/", "\ufeff", "/n", ",", and so on. Finally, these documents and their features (legal words) are represented in a form of vector space model, called bag of words (BOW) [17].

In this study, the experiments are conducted two different term-weighting schemes to justify the most suitable weighting scheme for our dataset. They are *tf-idf* and *tf-igm*. Each scheme can be described as follows.

tf-idf (term frequency-inverse document frequency) is a term-weighting scheme that is always used in information retrieval (IR) and text mining [18]. Its formula can be:

$$tf\text{-}idf_{t,d} = f_{t,d} \times \log\left(\frac{N}{df_t}\right) \tag{1}$$

where *tf* is frequency of term-word t occurs in a document d. Meanwhile, *idf* is the logarithmically scaled number of the total number of documents (N) in a corpus divided by the number of documents contain term-word t, denoted as *df*.

tf-igm (term frequency-inverse gravity moment) is the term-weighting scheme for text classification proposed by Chen *et al.* [19]. This scheme is a supervised term weighting (STW) used to measure the class distinguishing power of a term by combining term frequency with the *igm* measure. Its formula can be:

$$tf\text{-}igm_{t,d} = f_{t,d} \times (1 + \lambda \times igm(t_k)) \tag{2}$$

where $f_{t,d}$ is frequency of term t occurs in document d. For λ (Lambda), it is an adjustable coefficient that should be between 5.0 and 9.0. The purpose of the coefficient λ tries to maintain the relative balance between the global and local weights. Meanwhile *igm* is used to measure the inter-class distribution concentration of a term-word, which can be defined as follows.

$$igm(t_k) = \frac{f_{k1}}{\sum_{r=1}^{m} f_{kr} \cdot r} \tag{3}$$

where $igm(t_k)$ represents the *inverse gravity moment* of the inter-class distribution of term-word t_k, and $f_{kr}(r = 1, 2, ..., m)$ are the frequencies of t_k's occurring in different classes, which are in descending order with r being the rank. Mostly, the frequency f_{kr} refers to the *class-specific document frequency (df)* that is the number of documents containing the term-word t_k in the *r-th* class, denoted as df_{kr}.

4.2 Classifier Modeling of Each Civil Case Decision Component

After having the BOW from the stage of the civil case decision document pre-processing, it is passed to this stage, called as classifier modeling of each civil case decision component. To obtain the most appropriate classifier models, four supervised learning algorithms are compared. They are Naïve Bayes (NB), Support Vector Machine (SVM), and K-Nearest Neighbors (KNN).

Multinomial Naïve Bayes (MNB): The general term Naïve Bayes (NB) refers to strong independence assumptions in a classifier model [20–22], rather than the particular distribution of each feature. In this study, a feature is classified as a '*word*'. The NB model assumes that each feature that it uses is conditionally independent of another feature within the same class. The probability of features can be denoted as $f_1, f_2, ..., f_n$, that are assigned to the class c under the NB assumption as the following formula:

$$p(f_1, f_2, ..., f_n|c) = \prod_{i=1}^{n} p(f_1|c) \tag{4}$$

Thus, when using the NB model to classify an incoming example, the posterior probability is much simpler to work with and can be represented as:

$$p(c|f_1, ..., f_n) = p(c) \times p(f_1|c)...p(f_n|c) \tag{5}$$

The term MNB simply explains that each $p(f_i|c)$ is a multinomial distribution, rather than some other distribution. This works well for data that can easily be turned into counts, such as word counts in texts.

K-Nearest Neighbors (KNN): KNN is one of the simplest supervised machine learning algorithms. It can be used to perform classifications through the use of a majority voting mechanism. The training set is used as a decision rule to assign a class label to the incoming input under the K-closest consideration. In general, K should be a small odd number. Some-times, it can only be 1. Larger K values are generally not recommended although they are more accurate for the classification. This is because it takes longer to perform the task [23,24].

Support Vector Machines (SVM): SVM is an algorithm that can determine the best decision boundary between vectors that belong to a given class and those that do not. SVM is an extremely popular algorithm for text classification as it can return good results. SVM learns an optimal way to separate the training documents according to their class labels in n-dimensional space. The output of this algorithm is a hyperplane that maximizes the separation of data or feature vectors of documents belonging to different classes. Given an incoming document, SVM assigns a label based on the particular subspace to which its feature vector belongs to [25,26].

5 The Experimental Results

Recall (R), precision (P), and f-measure (F1) are four measures commonly used to evaluate the performance of prediction models [27]. After evaluating the obtained classifier models based on 10-fold cross validation, the average results of recall (R), precision (P), and F-measure (F1) are presented in Table 2.

Table 2. The evaluation of the classifier models with tf-idf

Component	MNB			SVM			KNN		
	R	P	F1	R	P	F1	R	P	F1
Dispute	0.79	0.89	0.84	0.24	0.58	0.34	0.73	0.38	0.50
Fact	0.77	0.87	0.82	0.21	0.53	0.30	0.11	0.24	0.15
Decision	0.80	0.85	0.82	0.96	0.34	0.50	0.64	0.49	0.56
Judgment	0.79	0.82	0.80	0.39	0.90	0.54	0.34	1.00	0.51
Average	0.79	0.86	0.82	0.45	0.58	0.42	0.45	0.52	0.43

Consider the experimental results in Tables 2 and 3. The *tf-igm* weighting scheme may be the most appropriate for this study since it is possible that the same '*word*' can be found in many classes but have differing importance in each class. Using *tf-igm* as the weighting scheme can present the particular importance of that '*word*' in a clear way.

In addition, the MNB classifiers can return the highest accuracy because the BOW representation matrix contains a large number of words. The MNB

Table 3. The evaluation of the classifier models with tf-igm

Component	MNB			SVM			KNN		
	R	P	F1	R	P	F1	R	P	F1
Dispute	0.89	0.91	0.90	0.77	0.79	0.78	0.78	0.82	0.80
Fact	0.84	0.89	0.86	0.75	0.78	0.76	0.75	0.80	0.77
Decision	1.00	0.96	0.98	0.77	0.76	0.80	0.80	0.85	0.82
Judgment	0.99	0.94	0.96	0.78	0.79	0.76	0.79	0.82	0.80
Average	0.93	0.92	0.92	0.77	0.78	0.77	0.78	0.82	0.80

algorithm works better with large numbers of features (words) as this algorithm implicitly treats all features as being independent of one another. Therefore, the high dimensionality problem that typically rears its head when dealing with high-dimensional data does not apply.

However, there is another challenge. Civil decision documents are prepared in a semi-structured format and contain legal language. Although they contain different components, they are still represented in the same format. Therefore, it is possible that some details in the components are similar. As an example, the "*fact*" component contains certain details that are explained in a similar way to the "*decision*" component. This makes it difficult to clearly classify these components. As a result, a feature selection technique is required. A new weighting scheme is also required that can specify the importance of each word in each specific component class to enhance classification performance.

As an experiment, we used 100 civil case decision documents that were different from the dataset used for classifier modeling. These documents were preprocessed and the *tf-igm* weighting was used. Finally, the MNB classifiers were employed to automatically identify the four major components in each document. The MNB classifier returned satisfactory results when working with real-world documents; however, the classifiers still require further enhancement. The results of identifying four major components in each document can be presented in Table 4.

Table 4. The results of identifying four major components in civil case decision document

Component	Recall	Precision	F1
Dispute	0.82	0.81	0.81
Fact	0.79	0.76	0.77
Decision	0.85	0.83	0.84
Judgment	0.84	0.81	0.82
Average	0.83	0.80	0.81

Consider the experimental results in Table 4. It can be seen that the MNB classifier can return satisfactory results although working with real-world documents. However, the classifiers still need to be enhanced.

6 Conclusion

This research aims to apply text classification techniques to identify major components in legal documents, especially civil case decision documents. The major components were defined as dispute, facts, decision, and judgment. It would be beneficial if we could recognize the major components in a civil case decision document before addressing other tasks such as text summarization and information extraction. To the best of our knowledge, this problem has never been addressed in previous studies. In our experiment setup, we compared two weighting schemes as *tf-idf* and *tf-igm* and three supervised machine learning algorithms as MNB, SVM, and KNN. Results indicated that the MNB classifiers returned the highest results when used with *tf-igm* weighting schemes. After testing by recall, precision and F1, satisfactory results were achieved by the MNB classifiers in identifying major components in Thai civil case decision documents at 0.83, 0.80 and 0.81, respectively.

References

1. Kowsrihawat, K., Vateekul, P.: An information extraction framework for legal documents: a case study of Thai supreme court verdicts. In: 2015 12th International Joint Conference on Computer Science and Software Engineering (JCSSE), pp. 275–280. IEEE (2015)
2. Palau, R.M., Moens, M.F.: Argumentation mining: the detection, classification and structure of arguments in text. In: Proceedings of the 12th International Conference on Artificial Intelligence and Law, pp. 98–107. ACM (2009)
3. de Maat, E., Krabben, K., Winkels, R., et al.: Machine learning versus knowledge based classification of legal texts. In: JURIX, pp. 87–96 (2010)
4. Boella, G., Di Caro, L., Humphreys, L.: Using classification to support legal knowledge engineers in the Eunomos legal document management system. In: Fifth International Workshop on Juris-Informatics (JURISIN) (2011)
5. Farzindar, A., Lapalme, G.: The use of thematic structure and concept identification for legal text summarization. Computational Linguistics in the North-East (CLiNE 2002), pp. 67–71 (2004)
6. Hachey, B., Grover, C.: A rhetorical status classifier for legal text summarisation. In: Text Summarization Branches Out (2004)
7. Teufel, S., Moens, M.: Summarizing scientific articles: experiments with relevance and rhetorical status. Comput. Linguist. **28**(4), 409–445 (2002)
8. Schilder, F., Molina-Salgado, H.: Evaluating a summarizer for legal text with a large text collection. In: 3rd Midwestern Computational Linguistics Colloquium (MCLC). CiteSeer (2006)
9. Galgani, F., Compton, P., Hoffmann, A.: Citation based summarisation of legal texts. In: Pacific Rim International Conference on Artificial Intelligence, pp. 40–52. Springer (2012)

10. Erkan, G., Radev, D.R.: LexRank: graph-based lexical centrality as salience in text summarization. J. Artif. Intell. Res. **22**, 457–479 (2004)
11. Timothy, D., Allison, T., Blair-goldensohn, S., Blitzer, J., Elebi, A., Dimitrov, S., Drabek, E., Hakim, A., Lam, W., Liu, D., et al.: MEAD a platform for multidocument multilingual text summarization. In: International Conference on Language Resources and Evaluation (2004)
12. Kumar, R., Raghuveer, K.: Legal document summarization using Latent Dirichlet Allocation. Int. J. Comput. Sci. Telecommun. **3**, 114–117 (2012)
13. Saravanan, M., Ravindran, B., Raman, S.: Improving legal document summarization using graphical models. Frontiers Artif. Intell. Appl. **152**, 51 (2006)
14. Sulea, O.M., Zampieri, M., Vela, M., Van Genabith, J.: Predicting the law area and decisions of French supreme court cases. arXiv preprint arXiv:1708.01681 (2017)
15. Howe, J.S.T., Khang, L.H., Chai, I.E.: Legal area classification: a comparative study of text classifiers on Singapore supreme court judgments. arXiv preprint arXiv:1904.06470 (2019)
16. Haruechaiyasak, C., Kongyoung, S., Dailey, M.: A comparative study on Thai word segmentation approaches. In: 2008 5th International Conference on Electrical Engineering/Electronics, Computer, Telecommunications and Information Technology, vol. 1, pp. 125–128. IEEE (2008)
17. Ko, Y.: A study of term weighting schemes using class information for text classification. In: Proceedings of the 35th International ACM SIGIR Conference on Research and Development in Information Retrieval, pp. 1029–1030. CiteSeer (2012)
18. Luaphol, B., Srikudkao, B., Kachai, T., Srikanjanapert, N., Polpinij, J., Bheganan, P.: Feature comparison for automatic bug report classification. In: International Conference on Computing and Information Technology, pp. 69–78. Springer (2019)
19. Chen, K., Zhang, Z., Long, J., Zhang, H.: Turning from TF-IDF to TF-IGM for term weighting in text classification. Expert Syst. Appl. **66**, 245–260 (2016)
20. Mohana, R., Sumathi, S.: Document classification using multinomial Naïve Bayesian classifier. Int. J. Sci. Eng. Technol. Res. (IJSETR) **3**(5), 1557–1563 (2014)
21. Xu, S., Li, Y., Wang, Z.: Bayesian multinomial Naïve Bayes classifier to text classification. In: Advanced Multimedia and Ubiquitous Engineering, pp. 347–352. Springer (2017)
22. Abbas, M., Memon, K.A., Jamali, A.A., Memon, S., Ahmed, A.: Multinomial Naive Bayes classification model for sentiment analysis. IJCSNS **19**(3), 62 (2019)
23. Beyer, K., Goldstein, J., Ramakrishnan, R., Shaft, U.: When is "nearest neighbor" meaningful? In: International Conference on Database Theory, pp. 217–235. Springer (1999)
24. Keller, J.M., Gray, M.R., Givens, J.A.: A fuzzy k-nearest neighbor algorithm. IEEE Trans. Syst. Man Cybern. **4**, 580–585 (1985)
25. Tian, Y., Ali, N., Lo, D., Hassan, A.E.: On the unreliability of bug severity data. Empirical Softw. Eng. **21**(6), 2298–2323 (2016)
26. Cristianini, N., Shawe-Taylor, J., et al.: An Introduction to Support Vector Machines and Other Kernel-Based Learning Methods. Cambridge University Press, Cambridge (2000)
27. Baeza-Yates, R., Ribeiro-Neto, B., et al.: Modern Information Retrieval, vol. 463. ACM Press, New York (1999)

Wearable Computing for Dementia Patients

Manasawee Kaenampornpan[1(✉)], Nguyen Duc Khai[2], and Khanabhorn Kawattikul[3]

[1] Polar Lab, Faculty of Informatics, Mahasarakham University, Khamriang Sub-district, Kantarawichai District 44150, Maha Sarakham, Thailand
manasawee.k@msu.ac.th

[2] Information and Communication Technology, ICT Lab, University of Science and Technology of Hanoi, 18 Hoang Quoc Viet, Cau Giay District, Hanoi, Vietnam
nguyenduckhai1998@gmail.com

[3] Department of Information Technology, Faculty of Social Science, Rajamangala University of Technology Tawan-Ok, Chanthaburi 22210, Thailand
khanabhorn.ka@gmail.com

Abstract. The number of dementia patients is increasing dramatically. As the dementia cannot be cured, only drug treatment can temporarily improve the symptoms. The doctor needs to adjust the medication according to the patients' symptoms. Therefore, the caregiver is required to give information about the patients' behavioral and psychological symptoms thoroughly in order for the doctor to adjust the medication efficiently. However, the caregiver has many daily tasks to complete. Thus, he might not know the behavioral and psychological symptoms in detail. This research proposes a wearable computing that monitor patient activities. This research is done based on the user centered design. Therefore, after the interview with doctors and caregivers, we finalized the activities that the doctors need to know in order to adjust the medication efficiently. The activities are stand-sit, shaking, walking, sitting and standing. Moreover, the best position of the wearable was concluded to be on the back of the patients. Then according to the previous works in fall detection systems, the models that we chose to compare their performance are feedforward neural network, support vector machine, decision tree and random forest. The selected features are Mean, Standard Deviation (STD), Peak counts, Zero crossing rate (ZCR), Spectral Energy and Spectral Entropy of x, y, z axis data from accelerometer and gyroscope. The result shows that the feedforward neural network achieved the highest accuracy.

Keywords: Dementia · Activity detection · Wearable computing · Activities monitoring · Accelerometer · Gyroscope · User centered design

1 Introduction

1.1 Dementia

Dementia causes a deterioration in cognitive function [1]. This affects memory, thinking, orientation, comprehension, calculation, learning capacity, language, and judgement.

P. Meesad and S. Sodsee (Eds.): IC^{2}IT 2020, AISC 1149, pp. 21–30, 2020.
https://doi.org/10.1007/978-3-030-44044-2_3

At the moment, there is no cure for dementia. But there are drug treatments that may temporarily improve the symptoms. The right treatment may change over time. Certain symptoms of behavioral and psychological symptoms of dementia (BPSD) may have a negative effect on patient's cooperativeness for receiving regular medical care. When the patient's symptoms change, the caregivers need the doctor to adjust the medication for the patient before they have higher levels of stress and depression from patient's behavior.

The number of elderly people with dementia is growing rapidly in an upcoming aging society era. According to the World health organization, there are around 50 million people that have dementia worldwide. There are also nearly 10 million new cases every year [1]. In Thailand most patients are living at home with their caregivers who are normally their spouse or relatives. However, the caregivers are not able to monitor them closely all of the time as they also have to do the housework or their income job. Moreover, at the moment, with the limitation on the number of available doctors, and costs of hospital visits, the caregivers can only take the patients to see the doctor every 1 to 3 months. Each visit the caregivers have to give information about patients' behavior and their emotions to the doctors for diagnosis. Due to the period of time between each visit, the caregivers might not be confident of their information. The inaccuracy of the information could lead to ineffectiveness of the doctors' medical adjustment for the patient.

2 Patient Activities Monitoring

There are many existing fall detection systems [2–8]. Many of them use accelerometers and gyroscopes. There are a few projects that use a smart watch too.

For example, Shen et al. [9] proposed a cloud computing based fall detection framework updating the model online with two-stage incremental update step. In the testing phase, 7 human subjects were asked to fall in 4 directions (front fall, back fall, left fall, and right fall) to a cushion. Waistbands are tied to subject's waist to collect the accelerometer and gyroscope data. After the experiment is done, precision and recall reached 0.937 and 0.935 in offline mode. For the cloud update, classification model can further increase the precision and recall to 0.991 (precision) and 0.990 (recall) after 300 iterations.

Ramachandran and Karuppiah mentioned machine learning has played a big part in fall detection system [10]. The existing works [2–8] either put the sensors on the wrist, chest, ankle or waist of the user to detect the fall. For example, Hussain et al. [11] tested data from accelerometer and gyroscope that are on the waist with SVM, KNN and random forest. They found that KNN gave the highest accuracy for fall detection.

In previous works, the activities that are detected limited to fall, walking, standing, lying down and sitting. However, in the dementia patients, the doctors are required to give information about the activeness/inactiveness of the patients that includes activity such as shaking as well.

3 Our Approach

3.1 Wearable Computing

Our project is based on the user centered design approach. After interviewing the doctors and caregivers, we found that the sensors cannot be put on the dementia patients where they can physically reach. This is because they will keep taking it off if the device is reachable. After our interview with the caregivers and doctors during the user requirement gathering stage, we found that the caregiver normally put the medication pad on the patient's back to avoid self-removing issues. Therefore based on this experience, we propose the position of the device to be worn in the similar position as the medication pad which is on the back of the patient as shown in Fig. 1 (Left).

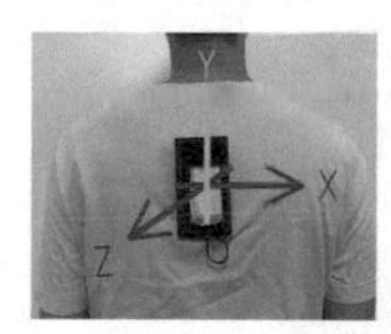

Fig. 1. Position of the wearable computing on the user's back (left) and wearable computing sensors (right).

As a result, the current device in use is a small, low-cost, lightweight and portable collection of sensors for monitoring patients' behavior as well as measuring data from patient physical's activity. Based on those measurements, the device quantifies patient's physical behavior, then send data to the database and triggers a messenger to call for a prediction from server. This is done with the help of installed components which are from right to left: FONA 808, ESP32 and LSM9DSO in Fig. 1 (Right). LSM9DSO is a sensor measuring accelerometer and gyroscope data from human gestures and ESP32 is used for connecting and transferring data to server. Unfortunately, FONA 808 component was not utilized in this project because we did not need to use a SIM card to connect to our local server. This component will be used in the future project.

3.2 Android Application for Data Collection

In this project, all the data measured from the wearable device will be stored in a database server. To control the flow of incoming data, a mobile application is created for labeling each data with the corresponding activity. After the focus group session with the doctors, we finalized the common activities that the doctors need to know in order to adjust the medication efficiently. The activities are stand-sit, shaking, walking, sitting and standing. As seen in Sect. 2, previous works do not detect an activity such as shaking before. This is another challenge for us. Figure 2 shows two interfaces of the android application for collecting the labelled data. First menu interface is the list of possible patient's activities that was shown for users to choose from. After choosing the activity for labeling, the application will switch to a second interface where users can start recording the data of chosen activities from sensors in an arbitrary amount of time.

Combining wearable device with mobile application, data collection phase is started by measuring accelerometer and gyroscope data from the patient. In some cases, the sensors capture abnormal data, such as high peak signals when patients stand up or a wave of continuous peaks when patients are walking or experiencing tremors. To be more precise in measuring activities according to the doctors' requirement mentioned before, the dataset is divided into two groups: active and inactive activities. The duration of the data collection process using wearable device is one week. Each day, we recorded each action for five times, measuring the sensors for 20 s each. The result we got is about 30,000 samples in total, which are then split into a train and test dataset.

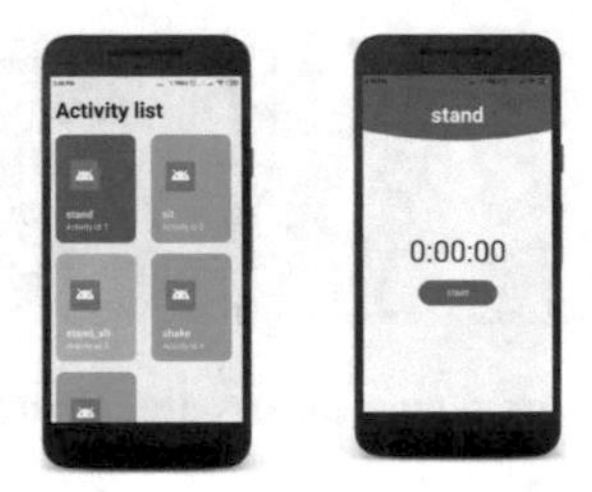

Fig. 2. GUIs of data collection application.

Inactive data consists of data points that are labeled with two activities: standing up and sitting down for a period of time. Physically, the main difference of two activities is the position of the subjects' bodies. The signals they produced have the same wave motions, but their amplitudes are different. As in the Figs. 3 and 4, y axis and z axis of accelerometer sensor is different from each other. The y axis of standing is higher than sitting down because of the device position. Z axis is lower as people tend to hunch their upper back when sitting down.

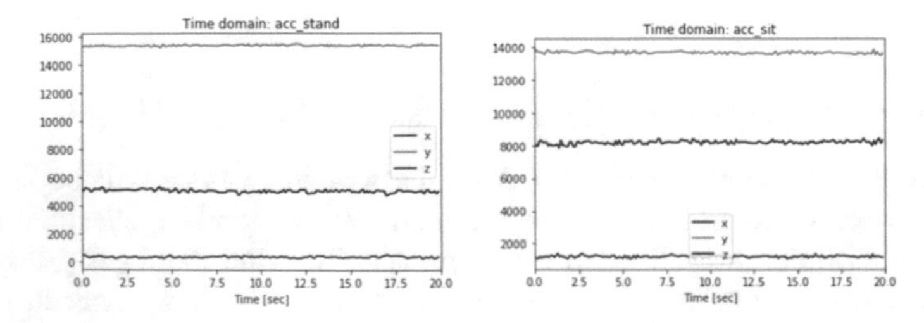

Fig. 3. Example of inactive data collected from the accelerometer for standing activity (left) and sitting activity (right).

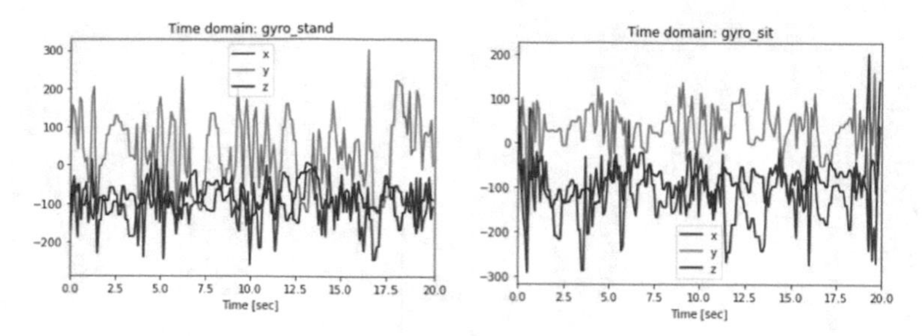

Fig. 4. Example of inactive data collected from the gyroscope for standing activity (left) and sitting activity (right).

On the other hand, the active data shows more fluctuating signals than inactive activities. The associated labels consist of walking, shaking and stand sit. For the shaking labelling, we collect data when user shakes their body back and forth resulting in a wave of high frequency on z and y accelerometer axis. We chose this activity because the doctor found that it is a common activity in the dementia patients. Walking activity is different in creating high frequency on x accelerometer axis and y gyroscope axis. Looking closely on other y and z axis in accelerometer, both axes have the same pattern but small waves of frequencies as shown in Figs. 5 and 6.

Lastly, the stand-sit activity refers to a moment when changing from stand up to sit down. Figure 7 is quite clear and straightforward in accelerometer sensor data, y axis rises up when standing up and z axis goes down when sitting down. With the differences in the signals of each activity, we are able to double-check if the labeling process is correct or not.

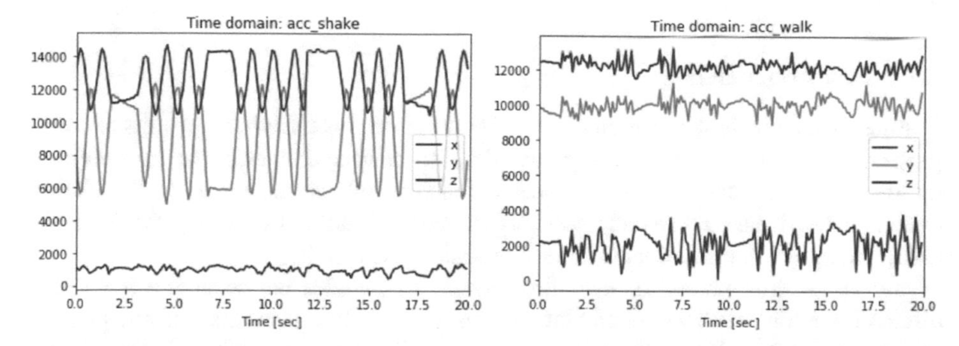

Fig. 5. Example of inactive data collected from the accelerometer for shaking activity (left) and walking activity (right).

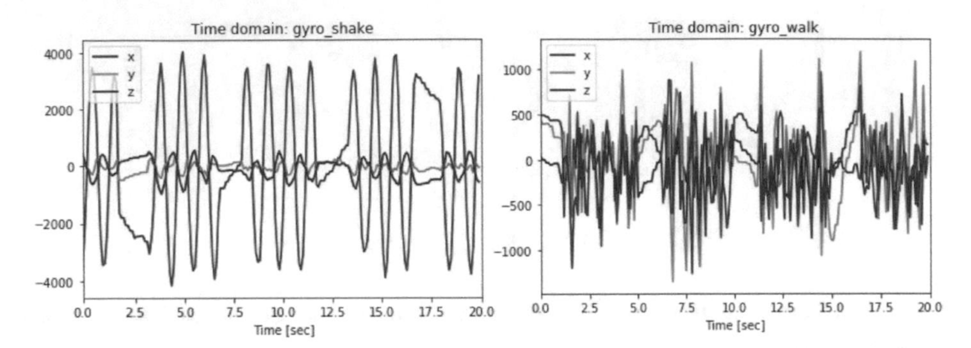

Fig. 6. Example of inactive data collected from the gyroscope for shaking activity (left) and walking activity (right).

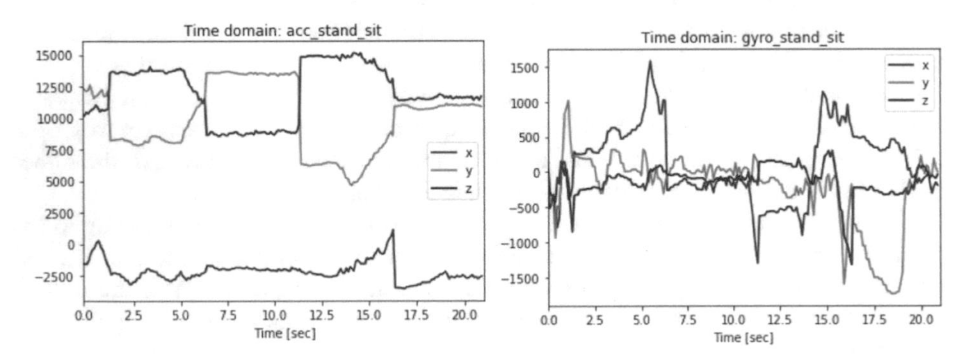

Fig. 7. Example of inactive data collected from the accelerometer (left) the gyroscope (right) for stand-sit activity.

3.3 Data Preprocessing

Feature extraction is the important step for most machine learning pipelines. In our project, this involves transforming raw data from accelerometer and gyroscope to multi-dimensional data. All these selected features will be the input for the classification model. Time domain and frequency domain will help us obtaining these features and gaining enough information to differentiate each activity.

In the labeling phase, the sampling rate is 10 samples per second. After testing different window sizes, we found that with smaller window sizes, less information will be extracted, which leads to wrong predictions. With a larger window, it costs us less input data for the future prediction phase. After testing each window size, we decided that a 2 s width was the optimal choice, giving enough samples to cover the moment of action we want to predict, as well as avoiding incorrect activity predictions that appear on the edge of one frame data ahead. As a result, after recording is finished, the signal is split into frames with a fixed window size of 2 s. Therefore, a single sequence data point in our dataset will have a length of 20 data. The data is then used to find the selected features. The selected features are Mean, Standard Deviation (STD), Peak counts, Zero crossing rate (ZCR), Spectral Energy and Spectral Entropy of x, y, z axis data from accelerometer and gyroscope. Before the training phase, all these features will be normalized. As a result, the input matrix size is 1×36.

3.4 Algorithms

Based on features we have extracted, several machine learning algorithms have been trained to predict which human activity is being performed. We then selected the best performing model to deploy on our system. The models that we experimented with are feedforward neural network, support vector machine, decision tree and random forest. These are the common models that were used in the previous fall detection systems mentioned in Sect. 2. In this section, we will discuss in detail each model and how we trained them.

Feedforward Neural Network. Feedforward neural network is a powerful tool to be considered in our system. However, since we have extracted a lot of features from a small amount of data samples, overfitting still occurred in the training phase. To overcome overfitting, we used two strategies: weight regularization and dropout [12]. Our neural network had 3 layers: the first layer has 64 units, the second has 32 units and the output layer has 5 units corresponding to 5 classes. Because we were dealing with a multi-class classification problem, soltmax function was applied in the output layer and for the rest, we used reLU. L2 regularization weight which was set to 1e4 and the dropout rate was 0.5. The network was trained for 200 epochs with a batch size of 64.

Support Vector Machine. Given labeled training data, the algorithm outputs an optimal hyperplane which will be used to categorize new examples. Nonetheless most practical datasets have non-linear relationships, so the decision boundary also needs to be non-linear. This problem can be overcome by using the kernel trick, which generates complex decision boundaries that can classify non-linear data in high dimensional space. There are two hyperparameters: a regularization parameter that reduces overfitting and gamma which defines how far the influence of a single training example reached. The kernel type we used will be poly regression with the degree of 3 and penalty parameter C set to 3.

Decision Tree. A decision tree includes nodes which represent decisions in order to predict test data points. It uses an entropy function to grow by estimating the purity of each classes. All impurity classes will be divided until there is none. As the dataset is relatively small but there are many features, our decision tree can easily overfit and create a complex decision tree. Other problems like unstability can occur: small variations in the data might result in a completely different tree being generated. By controlling the size of the decision tree, it will help us reduce overfitting by set max depth to 4. A very small number of leaves can also cause the tree to become unstable and overfit, whereas a large number will prevent the model from learning complex patterns. Setting minimum sample split to 2 and minimum sample leaf to 5 are our choices for the decision tree hyperparameter.

Random Forest. Random forest is a classification algorithm consisting of many decisions trees. It uses bagging and features randomness when building each individual tree to try to create an uncorrelated forest of trees, whose predictions by committee are more accurate than that of any individual tree. Each individual tree in the random forest splits out a class prediction and the class with the most votes becomes our model's prediction. Because the model simply makes use of numerous decision trees, most of the parameters will be the same. The number of trees is set to 100.

4 Results

The process of finding the best hyper-parameters is difficult and time consuming. Using grid search, we can perform the tuning of our parameter in order to determine the optimal values for a given model. Since we had to deal with an imbalanced data, accuracy as a metric can misleads the actual performance of the models. F1 score is a better metric in this case because it considers both the precision and the recall of the test to compute the score. With k-fold cross-validation of k = 5, the result with the highest accuracy and highest F1 score is the right hyper parameter we want to find.

From Table 1, Neural network achieved the highest accuracy and the highest F1 score. Other models' accuracy are mostly the same. The lowest accuracy belongs to the support vector classifier model, which is 0.9563 but their F1 score are similar to the random forest model.

On the test set shown in Table 2, Neural network has the highest F1 score and accuracy. Random forest has the lowest accuracy score but its F1 score is higher than the one from decision tree. To further analyze the model's performance, we plotted a confusion matrix to see the network's predictions on the test set in Fig. 8. Sitting class have the best score and the lowest belong to standing class. Others like shaking and walking classes, both of them are similar to each other with about 0.8 accuracy score. Stand-sit class is different which can be misclassified to all other activities, making it the second lowest score out of five.

Table 1. Cross-validation results.

Models	Accuracy	F1 score
Support vector classifier	0.9563	0.9628
Decision tree	0.9565	0.9574
Random forest	0.9613	0.9628
Neural network	0.9818	0.9897

Table 2. Model's performance on the test dataset.

Models	Accuracy	F1 score
Random forest classifier	0.6392	0.6656
Decision tree classifier	0.6896	0.5606
Support vector classifier	0.7506	0.7585
Neural network	0.7877	0.7997

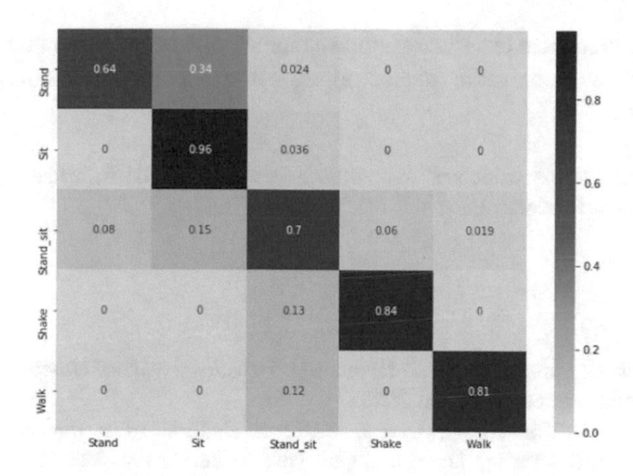

Fig. 8. Confusion matrix heat map of neural network.

5 Conclusion and Future Works

From the results, we observed that the neural network is the best model out of four classifiers. Nevertheless, the accuracy does not achieve our expected result. This is due to the limitation of dataset and invariability of the activity.

Looking back to the neural network confusion matrix, inactive activities have the highest and the lowest accuracy score. Standing and sitting are two noticeable activities but the difference in amplitude is low which may confuse the model. For active activities, three of them have a decent score, where Stand-sit has the lowest. This activity is the combination of both active and inactive, which is a possible explanation for the errors.

Before the training phase started, we tried using a low pass filter to remove noise from the gyroscope data. Theoretically, this could help improving the model. However, things did not follow our expectations, as the models performed worse afterwards. Perhaps by filtering out high frequency components, we had discarded some important information which is necessary to correctly classify the signal.

In this part of the project, as we design it based on the user centered design approach, our main aim is to gather the user requirement from the doctors and caregivers. In order to have the same understanding with the doctors and caregivers who have different background, the prototype is quickly created to be used as a tool to communicate with the doctors and caregivers. We have built a prototype system that consists of an IoT device, server and android application. To interact with the devices, we created a server for transferring and storing data from the IoT device. The mobile application was created for labeling data which used to train the neural network model and test them with the accuracy score of 0.787. The android application had another function, which is to stream data and show the prediction results. Our next step of the project is to increase the efficiency of the prediction model.

We believe that there are several improvements can be made to this project such as collect more data from more users, optimized the feature selection and machine learning models.

Acknowledgments. This work was supported by the Faculty of Informatics, Mahasarakham University and Thai Research Fund.

References

1. World Health Organization: WHO Homepage. https://www.who.int/news-room/fact-sheets/detail/dementia. Accessed 14 Jan 2020
2. Bourke, A.K., O'Brien, J.V., Lyons, G.M.: Evaluation of a threshold-based tri-axial accelerometer fall detection algorithm. Gait Posture **26**(2), 194–199 (2007)
3. Wearable Technologies: WT Homepage. https://www.wearable-technologies.com/2019/07/the-5-best-fall-detection-wearables-in-2019. Accessed 14 Jan 2020
4. Guo, H.W., Hsieh, Y., Huang, Y,. Chien, J,. Haraikawa, K,. Shieh, J.: A threshold-based algorithm of fall detection using a wearable device with tri-axial accelerometer and gyroscope. In: 2015 International Conference on Intelligent Informatics and Biomedical Sciences (ICIIB), pp. 54–57. IEEE, Japan (2015)
5. Pang, I., Okubo, Y., Sturnieks, D., Lord, S., Brodie, M.A.: Detection of near falls using wearable devices. J. Geriatr. Phys. Ther. **42**(1), 48–56 (2019)
6. Sanchez, J.A.U., Muñoz, D.M.: Fall detection using accelerometer on the user's wrist and artificial neural networks. In: XXVI Brazilian Congress on Biomedical Engineering, pp. 641–647. Springer, Singapore (2019)
7. Chandra, I., Sivakumar, N., Gokulnath, C.B., Parthasarathy, P.: IoT based fall detection and ambient assisted system for the elderly. Cluster Comput. **22**(1), 2517–2525 (2019)
8. Ranakoti, S., Arora, S., Chaudhary, S., Beetan, S., Sandhu, A., Khandnor, P., Saini, P.: Human fall detection system over IMU sensors using triaxial accelerometer. In: Computational Intelligence: Theories, Applications and Future Directions, vol. I, pp. 495–507. Springer, Singapore (2019)
9. Shen, J., Chen, Y., Shen, Z., Liu, S.: A two-stage incremental update method for fall detection with wearable device. In: 2018 IEEE SmartWorld, Ubiquitous Intelligence & Computing, Advanced & Trusted Computing, Scalable Computing & Communications, Cloud & Big Data Computing, Internet of People and Smart City Innovation (SmartWorld/SCALCOM/UIC/ATC/CBDCom/IOP/SCI), pp. 364–371. IEEE, China (2018)
10. Ramachandran, A., Karuppiah, A.: A survey on recent advances in wearable fall detection systems. Biomed. Res. Int. **2020**, 1–17 (2020). (Article ID 2167160)
11. Hussain, F., Hussain, F., Ehatisham-ul-Haq, M., Azam, A.A.: Activity-aware fall detection and recognition based on wearable sensors. IEEE Sens. J. **19**(12), 4528–4536 (2019)
12. Srivastava, N., Hinton, G., Krizhevsky, A., Sutskever, I., Salakhutdinov, R.: Dropout: a simple way to prevent neural networks from overfitting. J. Mach. Learn. Res. **15**(Jun), 1929–1958 (2014)

Supplement Products Data Extraction and Classification Using Web Mining

Nantawat Thongmaun(✉) and Wachirawut Thamviset

Department of Computer Science, Faculty of Science, Khon Kaen University, Khon Kaen 40002, Thailand
nantawatt@kkumail.com, twachi@kku.ac.th

Abstract. Currently, many product sellers like to advertise their supplement products on web. However, there are some ads showing messages to deceive consumers. This work presents a system to extraction supplement products advertisement data from web and classifies the illegal ads that show misleading properties. Therefore, we proposed a method to automatic search and extract ads text from multiple websites using defined supplements keywords. Then, the extracted ads texts were preprocessed by word segmentation, stop words eliminate methods, and classified by the misleadingness words database that be prohibited by the Food and Drug Administration of Thailand. All illegal classified ads would be computed TF-IDF vectors and stored in an illegal reference database. However, some illegal ads avoided to use the prohibited words that they can be classified as legal. Therefore, they would be re-classified by measuring the similarity with all ads in the reference database. The experimental results show that the proposed system can detect forbidden ads with an accuracy of 0.775.

Keywords: Supplement products · Classification · Web mining

1 Introduction

Food supplements have become a popular choice for people who wish to take care of their health. They can be purchased in department stores, shops, and through direct sales on the internet. Food supplements are classified as food, not medicine, and therefore have no properties capable of preventing or treating diseases. Statements involving the treatment of disease, weight loss, etc., should, therefore, be ignored. Their objective, therefore, is only to improve one's health [1].

However, false advertising, erroneous claims, and exaggerated properties remain a problem. Correct and proven data is difficult to obtain due to the many sources available and the time required to collect sufficient information. As a result, we used data extraction for product supplementation on the web, which separated the HTML document structure, dividing the blocks of information. We then sorted the ads using a list of words prohibited by the Food and Drug Administration of [2] in advertising. Misleading advertising [3] at this stage was used as a reference for the analysis of classified ads. As sellers may try to avoid using such words in their ads, we employed word segmentation using the longest matching method. Word elimination discontinues

P. Meesad and S. Sodsee (Eds.): IC^{2}IT 2020, AISC 1149, pp. 31–39, 2020.
https://doi.org/10.1007/978-3-030-44044-2_4

the use of words that do not affect the meaning of a sentence and uses synonyms to reduce features through the TF-IDF vector. Additionally, cosine similarity analysis compared the similarities between documents.

This paper describes the classification advertisement of food supplement products as follows: Sect. 1 provides an introduction to the topic with a brief statement of our intent. Section 2 describes previous works related to our study, and Sect. 3 presents our framework and methodology. Section 4 shows our experiment results, and lastly, Sect. 5 presents our conclusion.

2 Related Work

Liao et al. [4] proposed methods for extracting data from 96 stores in the UNIMALL and ESCA. Their data extraction, in Japanese, divided the work into two parts: the first being the division of blog blocks, and the second involved tagging to specify the information. Similar to our work, the division of the webpage was analyzed and divided according to the HTML document structure through the use of a machine learning technique within the SVM (support vector machine).

Kovacevic et al. [5] proposed a method to divide the block according to the page layout format: top (header), bottom (footer), and left, right, and center; creating a tree structure, referred to as the M (method) - Tree.

Saipech and Seresangtakul [6] proposed an automatic scoring system to answer examination questions by measuring similarities with cosine. Student answers in short Thai sentences were segmented with Thai words using the longest matching algorithm and dictionary (Lexitron®). While most systems do not use synonyms, the results of the two comparative systems found that the system that contained synonyms was closer to that of an expert than the systems that did not use synonyms.

Rababah and Al-Taani [7] presented a short, automatic Arabic essay rating using a cosine similarity measurement method between student responses and model answers, which used rooting and synonyms to obtain the correct answer, which was then converted to a vector using the TF-IDF method. The researcher used a single set of data, in which 11 questions were answered by 50 students. The average length of the students' answers was four sentences, 50 words, or 200 characters. The measured similarity between the student answers and the model answers was 95.4%.

Lahitani et al. [8] presented an online essay evaluation scoring system in the Indonesian language, using the TF-IDF weighting method similar to that of cosine at the word level. Some documents were preprocessed, in which the tokenizing process incorporated the Nazief-Adriani algorithm which measured the similarities in the ranking model.

Viriyavisuthisakul et al. [9] presented the classification of social media messages using the Thai text website PANTIP to classify messages into four classes, which were defined by experts. Their research contained token data using stop words, KUCUT for word segmentation, and two feature extraction methods, TF-IDF and TF, which compared ten distance measurements. The results demonstrated that the Bray-Curtis distance of the TF-IDF was suitable for the classification of Thai social messages.

Dhar et al. [10] proposed two ways of determining similarities: cosine similarity and Euclidean distance. Five types of domains from news websites (in the Bangladesh language) were compared. Over 1000 documents were divided into 50% trend and 50% test in order to extract the properties using the TF-IDF vector.

2.1 Web Data Extraction

Web data extraction techniques process large amounts of unstructured content on the web [11] and capable of extracting and copying data more accurate than humans [12].

2.2 Word Segmentation

Thai word segmentation without markers for word boundaries differs greatly from that for English words, in which two techniques were proposed: dictionary-based (DCB) and machine learning-based (MLB) [13]. DCB is a word segmentation technique that matches words from a dictionary; however, this method is weak in word segmentation, as words may be ambiguous or simply not found in the dictionary. MLB built statistical models are that which are evaluated from corpus training using machine learning techniques. In this research, we used the DCB word segmentation technique and applied the longest matching algorithm via the Lexitron® dictionary [14].

2.3 Term Frequency – Inverse Document Frequency (TF-IDF)

TF-IDF is a statistical method that evaluates the importance of words to a commonly used vector space model, whereas Term Frequency (TF) is the frequency of words appearing in each document. Inverse Document Frequency (IDF) refers to the inverse of the document frequency by reducing the weight of frequently used words, increasing the weight of rare words, through the following equations:

$$tfidf_{i,j} = tf_{i,j} \times idf_i \tag{1}$$

$$tf_{i,j} = \frac{n_{i,j}}{\sum_k n_{i,j}} \tag{2}$$

Where $n_{i,j}$ the number of occurrences of I in document j is, $\sum_k n_{i,j}$ is the sum of the occurrences of all terms in the document.

$$idf_i = log_e \frac{d}{df_i} \tag{3}$$

Where d is the total number of document sets, df_j is the total number of documents containing the same term.

2.4 Cosine Similarity

Cosine similarity measures the similarity between two vectors. This method can be used to obtain the similarity scores between two sentences or documents and is often combined with the TF-IDF. The cosine is between 0 and 1; in which the higher the cosine, the more similarity between the two sentences or documents. The formula for cosine similarity is as follows:

$$cos(RA, AA) = \frac{\sum_{j=1}^{t} W_j d_j}{\sqrt{\sum_{j=1}^{t} (w_j)^2} \times \sqrt{\sum_{j=1}^{t} (d_j)^2}} \quad (4)$$

Where W is the word frequency vector of document W and d is the word frequency vector of document d.

3 Methodology

In our work, we present an ad classification for food supplements that boasts exaggerated properties by dividing two main parts: the data extraction process and the classification process. In the first part, data is extracted from the advertisement of supplemental products. The second part, preprocessing data, begins with the classification of false advertising words as a reference for comparison. Words are converted to vector form TF-IDF and the cosine similarities are compared. Lastly, the advertisements are checked by experts. An outline of our methodology is presented in Fig. 1.

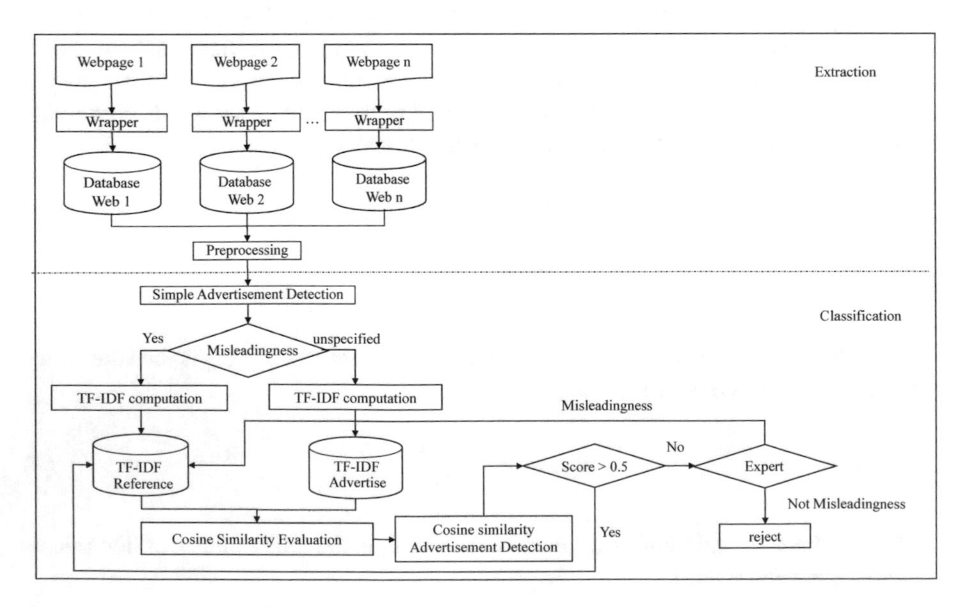

Fig. 1. Overview of research phases.

3.1 Extraction Stage

This section prepares the advertisement of supplements through the transformation into vectors in which to determine similarities through the following steps:

Web Extraction. In this process, information is extracted using relevant keywords, by looking at the structure of the DOM tree (Document Object Model) and extracting the information from the advertising text. We use web extraction tools, Beautiful Soup [15]. The Python library for extracting data from HTML and XML files works with the parser to provide methods for navigating, searching and editing trees for parsing.

In the extraction of a website advertise supplement products. We only extract text. We analyze from the HTML tree structure in which the text is placed in the tag <div> down until we find the tags <p>, <u> and <li>.

Preprocessing. At this stage, information is prepared for use, in which the main objective is to obtain important features or keywords from a data set. Preprocessing includes the removal of HTML tags, symbols, word segmentation, and synonym conversion.

3.2 Classification Stage

Simple Advertisements Detection. At this stage, we will detect any false or misleading information by using the prohibited word database, containing 198 words, provided by the Food and Drug Administration [16]. The results of this classification advertisements and were divided into two types. The first type is a system for detecting misleading statements, which, in this section, we will use as reference answers.

In the second type is unspecified answer (unspecified answer is an advertisement that cannot be classified by an advertisement detection process, which has false by-word avoidance and that is not wrong.), which we will find the similarities between the two types in the next section. The experts (nurses and pharmacists) label the advertisements as either right or wrong. Nurses and pharmacists are empowered under the Food Act 1979 revised version 2562 and use the Food and Drug Administration announcement. Subject: Food advertising guidelines 2018 in the classification of dietary supplement advertising [2, 17] (Table 1).

Table 1. Example of prohibited word.

ID	prohibited word
1	ป้องกันโรคหัวใจ (prevent heart disease)
2	ลดน้ำหนัก (lose weight)
3	ดักจับไขมัน (trap fat)
4	รักษาโรคมะเร็ง (Prevent cancer)
5	ไร้ผลข้างเคียง (no side effect)

TF-IDF Computation. This step converts the word to vector format. The word will be processed by TF-IDF, in which the weight of the word is obtained by comparing the questions with the importance of the words in the word bag. The number of keywords found in the question is then counted. The word frequency determines their weight. The TF-IDF is then stored in the database.

Cosine Similarity Evaluation. This step uses cosine similarity to estimate the similarity between a misleading answer and an unspecified answer. The TF-IDF vector performs the comparison and similarities are given in the range between 0 and 1.

Cosine Similarity Advertisement Detection. A value close to one indicates the use of misleading text, whereas values close zero are labeled as not misleading. A similarity of 0.5 requires no expert response, whereas similarities below 0.5 will be sent for verification by the expert system. If the expert replies 'wrong' it will be sent to the TF-IDF reference database. If it is not misleading (right), it will be removed immediately to prevent duplication.

4 Experimental and Result

4.1 Data Preparation

Data obtained from the extraction of food supplement advertisements in five websites in over 683 products were divided as follows: Website A contained 183 products, Website B contained 22 products, Website C contained 133 products, Website D contained 42 products, and Website E contained 303 products. In the advertisement detection section, the following results were determined: Misleading: 351; Unspecified: 351.

4.2 Performance Evaluation

Performance evaluation was determined through the measurement of Precision, Recall, F-Measure, and Accuracy as follows:

$$Precision = \frac{TP}{TP + FP} \tag{5}$$

$$Recall = \frac{TP}{TP + FN} \tag{6}$$

$$F\text{-}Measure = \frac{2 \times Precision \times Recall}{Precision + Recall} \tag{7}$$

$$Accuracy = \frac{TP + TN}{TP + FP + FN + TN} \tag{8}$$

Where Precision is the probability that the system predicts misleading advertisements, Recall is the probability that the system will detect misleading advertisements from the total number of advertisements, F-measure is the mean value of Precision and Recall, and Accuracy is the system's ability to accurately predict the results. TP is the number of advertisements that the system answers and the experts determine to be misleading. FP is the number of advertisements that the answering system claims are misleading, and where expert answers are not misleading. TN is the number of advertisements that the answering system clams are not misleading, and the expert answers are misleading. FN is the number of advertises where both the answering system and expert answers are not misleading.

In the advertisement detection section, the system detected 351 of 531 deceptive advertisements. The similarities with that of the expert examinations ranged from 0.3–0.9, as indicated in Tables 2 and 3, below.

Table 2. Calculation result of advertise.

Similarity	TP	FP	FN	TN	Total
0.3	410	32	121	119	682
0.4	374	14	157	137	682
0.5	354	0	177	151	682
0.6	354	0	177	151	682
0.7	352	0	179	151	682
0.8	352	0	179	151	682
0.9	351	0	180	151	682

Table 3. Calculation result of advertise.

Similarity	Precision	Recall	F-measure	Accuracy
0.3	0.927602	0.772128	0.842754	0.77566
0.4	0.963918	0.704331	0.813928	0.749267
0.5	1	0.666667	0.8	0.740469
0.6	1	0.666667	0.8	0.740469
0.7	1	0.6629	0.797282	0.737537
0.8	1	0.6629	0.797282	0.737537
0.9	1	0.661017	0.795918	0.73607

From Table 2, the similarity value 0.3 indicates the most accurate classification of misleading advertising. The similarity value 0.5 states 'not to release' an advertisement that the system claims are wrong, but the experts say is right. In Table 3, the similarity values of 0.3 are very accurate, in which the Accuracy, F-measure, and Recall were greatest at 0.775, 0.842, and 0.772; respectively.

5 Conclusion

This article presents the extraction of information from supplement product websites and automatically classifies hyperbolic nutritional properties using a similar method of cosine. Illegal advertisements, both direct and false by-word avoidance, are detected via the use of a database of prohibited words in the advertisement of dietary supplements from the Food and Drug Administration of Thailand. The direct false advertisements acted as a reference in order to compare the omission of false advertising as monitored by experts. However, in some cases, we noticed that the analogy comparison often demonstrated low similarity because the sentences had very different lengths, which led to a decrease in system performance. In future work, we intend to add a set of food ads, as well as complement, semantic, and contextual analysis which we feel will further improve performance.

Acknowledgment. We thank you to the pharmacists and nurses of King Taksin Hospital. For labeling of supplement products.

References

1. Food and Drug Administration. http://www.fda.moph.go.th. Accessed 10 Oct 2019
2. Bureau of Food: Announcement of the Ministry of Public Health No. 275 - 300 (Issue No. 293, 2005, regarding dietary supplements) (2005). http://food.fda.moph.go.th/law/data/announ_fda/61_advertise.PDF. Accessed 10 Oct 2019
3. Russo, J., Metcalf, B., Stephens, D.: Identifying misleading advertising. J. Consum. Res. **8**, 119–131 (1981). https://doi.org/10.1086/208848
4. Liao, C., Hiroi, K., Kaji, K., Kawaguchi, N.: An event data extraction method based on HTML structure analysis and machine learning. In: 2015 IEEE 39th Annual Computer Software and Applications Conference, pp. 217–222. IEEE, Taichung (2015)
5. Kovacevic, M., Diligenti, M., Gori, M., Milutinovic, V.: Recognition of common areas in a web page using visual information: a possible application in a page classification. In: 2002 Proceedings of the IEEE International Conference on Data Mining, pp. 250–257 (2002)
6. Saipech, P., Seresangtakul, P.: Automatic Thai subjective examination using cosine similarity. In: 2018 5th International Conference on Advanced Informatics: Concept Theory and Applications (ICAICTA), pp. 214–218 (2018)
7. Rababah, H., Al-Taani, A.T.: An automated scoring approach for Arabic short answers essay questions. In: 2017 8th International Conference on Information Technology (ICIT), pp. 697–702 (2017)
8. Lahitani, A.R., Permanasari, A.E., Setiawan, N.A.: Cosine similarity to determine similarity measure: study case in online essay assessment. In: 2016 4th International Conference on Cyber and IT Service Management, pp. 1–6 (2016)
9. Viriyavisuthisakul, S., Sanguansat, P., Charnkeitkong, P., Haruechaiyasak, C.: A comparison of similarity measures for online social media Thai text classification. In: 2015 12th International Conference on Electrical Engineering/Electronics, Computer, Telecommunications and Information Technology (ECTI-CON), pp. 1–6 (2015)

10. Dhar, A., Dash, N., Roy, K.: Classification of text documents through distance measurement: an experiment with multi-domain Bangla text documents. In: 2017 3rd International Conference on Advances in Computing, Communication Automation (ICACCA) (Fall), pp. 1–6 (2017)
11. Ferrara, E., De Meo, P., Fiumara, G., Baumgartner, R.: Web data extraction, applications and techniques: a survey. Knowl.-Based Syst. **70**, 301–323 (2014). https://doi.org/10.1016/j.knosys.2014.07.007
12. Vanden Broucke, S., Baesens, B.: Practical Web Scraping for Data Science: Best Practices and Examples with Python. Apress, New York (2018)
13. Haruechaiyasak, C., Kongyoung, S., Dailey, M.: A comparative study on Thai word segmentation approaches. In: 2008 5th International Conference on Electrical Engineering/Electronics, Computer, Telecommunications and Information Technology, pp. 125–128 (2008)
14. GitHub: PyThaiNLP/pythainlp: Thai Natural Language Processing in Python. https://github.com/PyThaiNLP/pythainlp. Accessed 24 Oct 2019
15. Beautiful Soup Documentation — Beautiful Soup 4.4.0 documentation. https://www.crummy.com/software/BeautifulSoup/bs4/doc. Accessed 24 Oct 2019
16. Food and Drug Administration: Notice of the Food and Drug Administration Re: Food Advertising Regulations B.E. 2561 (2018). http://newsser.fda.moph.go.th/food/LawNotification%20of%20Ministry%20of%20PublicHealth06.php. Accessed 12 Oct 2019
17. Food Act 2522 together with Ministerial Regulations and the announcement of the Ministry of Public Health (revised version 2019). http://www.fda.moph.go.th/sites/food/law1/food_law.pdf. Accessed 30 Oct 2019

Aquaponics Systems Using Internet of Things

Patcharapol Boonrawd, Siranee Nuchitprasitchai(✉), and Yuenyong Nilsiam

King Mongkut's University of Technology North Bangkok, Bangkok, Thailand
s6107011858626@email.kmutnb.ac.th,
siranee.n@it.kmutnb.ac.th, yuenyong.n@eng.ac.th

Abstract. The waste from fish causes the quality of the water to be lower or even harmful for aquatic animals. The nitrate in the water can be removed using plants as doing in hydroponics system. The waste water from the fish pond cycling through the hydroponics farm to removal the nitrate and back to pond. Combining fish tank system and hydroponics farm, it is aquaponics system. In this research, the proposed system using Internet of Things concept (IoT) to help monitoring and controlling the system. Arduino microcontroller board is used as the main controlling and processing. All sensors are connected to the board and data is transferred to the database on the cloud via wireless Internet connection. The lights, the fish feeder and the pumps are automatically controlled by the software on the Arduino board. However, the lights can be controlled manually using an application on a smartphone. The application shows humidity, temperature, moisture data from the sensors and the status of the lights. The proposed system was developed successfully and the result shows that the system is useable.

Keywords: Aquaponics · Internet of Things · Hydroponics

1 Introduction

Nowadays, a cultivation of aquatic animal has been growing enormously. It has been developed into an industry of raising aquatic animals for local consumption and exporting. It caused a problem of low quality of water due to accumulation of waste from fish in the pond. They are animal waste and decomposition of animal feed which cause aquatic animals in need of more oxygen and risk for infecting with a disease. Besides, the quality of water in the pond, draining waste water with nitrogen into natural sources of water will cause a eutrophication. The eutrophication is a situation that aquatic plant and microalgae rapidly grow causing low oxygen in the water and water turn into wastewater. The closed system for aquaculture that has water treatment and cycling the water efficiently is not only maintaining the quality of the water in the system, it is environmentally friendly as well. The method is favor for controlling the waste amount in the water in order to control the number of bacteria for complete bioreaction. Using plant for organic matter removal [1], especially nitrogen and phosphorus, is a popular method since the process of treatment uses low energy and environmentally friendly. This has become a concept of feeding aquatic animals and planting together, aquaponics. The combination of feeding aquatic animals and soilless

P. Meesad and S. Sodsee (Eds.): IC²IT 2020, AISC 1149, pp. 40–48, 2020.
https://doi.org/10.1007/978-3-030-44044-2_5

planting (hydroponics) is proposed. The main idea is that from the pond of aquatic animals which is full of nitrogen compound, ammonia or nitrate. This compound is macronutrient for plant for its growing. Therefore, the water in the pond is treated and the plant can be sold for additional income. However, applying aquaponics idea successfully depends on the proportion of aquatic animals and hydroponics. The waste from the aquatic animals need to be enough for the plants to grow, otherwise, the plants would not have enough nutrients. On the other hand, if the waste from the pond is excessively for being used by the plants, the accumulation of the waste in the pond would be too high for the aquatic animals.

An application of Internet of Things (IoT) [2] is for aquaponics management using an interface on a computer or a smartphone to control the system via wired or wireless Internet. For the convenience of usage and connecting to the system, an IoT cloud platform would be used for database and system controlling. The main controlling system is using a microcontroller board. The board is like a small computer that capabilities of connecting to a monitor and keyboard supporting computer programming or general uses, such as, word processing, spreadsheet, Internet surfing, emailing, etc. The board can be programmed to control electronic devices. For example, controlling relay to turn on and turn off a pump based on setting times [3]. Anyway, security is a concern for the use of the board [4].

According to the above information, a system with the combination of hydroponics, aquaculture, and IoT is proposed in this research. Hopefully, the system would be an alternative for personal growing vegetable and feeding fish together. The result is the user has clean vegetable and fish without sewage discharging into the natural sources of water.

2 Literature Review

Smart farm is using modern technology to help farmers solve many of their problems. The new concept is the high accuracy of data for agricultural. It is a strategy for famer to be environmentally friendly by adapting resources usage based on the condition of the land. This can be applied to the farm for both animal and plant. The smart farm focuses on optimizing resources usage based on data from sensors, so it is really meet the need of animal and plant [5]. Smart garden is a similar system to manage everything in the garden automatically [6]. The system collects environment data, process and control the condition without human involve. The system can be controlled remotely for a far which gives users comfort and easy to work. The temperature and humidity are stored and used to control the watering system via browser. The information is shown on the web browser. Automatic watering system works as programmed according to the data from wireless sensor network [7]. The system is in good efficiency based on the monitoring of humidity in the soil using sensors. A problem is that the sensor is not working if the humidity is greater than 90%. Small drone can be used to capture pictures of the area for land management. A simple weather station can also be used to record the data and the data is analyzed for management planning. Farmers would satisfy greatly with the information they have received [8].

Hydroponics [9] is a cropping method which is popular recently. It eliminates many limitations, area, size, and cost. The technique is suitable for almost all kinds of plants, vegetable, fruit tree, flowerer, garden tree, climbing plant, and perennial plant. However, it is very popular for fast growing plants and can be harvested in a short time. Hydroponics can be used to avoid many problems, saline soil, acid soil, and improper weather. Hydroponics for salad greens in condominium or rental room is managed by Arduino in real-time [10]. The sunshine is measured by sensor to control the LED light for the substitution. The water level in the system is monitor by the sensor and the system adjust the water level in real-time based on the data. The Blynk application is used to support controlling the light and water outlet in the system. The result of the research shows that vegetables in the system grow faster than the traditional method about 10 days and the leaves are bigger too. The main reason is the stable of light intensity that was controlled by the system is variate only 8.83% compare to 21.66% of the normal method.

Combining a fish tank and a hydroponics farm together is called an aquaponics. The water circulation system of aquaponics is bringing water from the fish pond to the hydroponics and back. The system consists of two main parts which are the fish pond and the hydroponics plant. After feeding, the fish excrete ammonia which is highly toxic to fish. Ammonia in the water with oxygen, nitrifying bacteria would change the ammonia into nitride and then nitrate, respectively. While nitrate is toxic to fish, it is a good nutrient to plant. If the nitrate water is used to grow plant, the plant would absorb the nitrate which improve the quality of the water. The water would flow back to the pond with good quality for the fish. Because it is a closed loop water system, the waste from the fish is cleaned out by the plant and keep circulating. The system is appropriate for growing organic vegetables. It is safe for producer, consumer, and environment [11].

The problem of raising aquatic animals is how to remove the nitrate from their excretion in the water. Nitrate treatment system was created in a research with a 50-m long PVC and 2.5 cm diameter tube, filled in with 2,860 plastic balls, and adding methanol 5% at 10 milliliters per hour. The result shows that the nitrate treatment system using a long tube has a good efficiency for raising tiger prawn breeder in a building [12]. However, another research [13] proposed aquaponics system. Wastewater from the fish tank is transferred to a hydroponics farm. Plant uses the waste as its sources of nutrient to grow. After that the water is transferred back to the fish tank as clean water for fish. The hydroponics is used instead of the nitrate treatment system. Combining the fish tank and the hydroponics farm has created a good ecosystem. The proportion of plant to fish in aquaponics system has direct effect to the efficiency of the system [14]. The study showed a strong relationship between number of tilapias and sweet turnips in aquaponics. The result shows that the increasing number of plant has improved the efficiency of the system. More plant means ammonia and nitrate will be removed more from the water which make the quality of the water better and the fish grow quicker. The general rule is that the balance of nitrate generated by fish and removed by plant [15].

For a system to be monitored and controlled, technology called Internet of Things (IoT) is playing an important role. IoT is an environment that things are connected and communicated to each other through protocols via both wired and wireless network [16]. Things are uniquely identified from each other, able to sense surrounding environment, and interactive and collaborative. The ability to communicate of these things would bring us new innovations and services. For example, sensors in a house detects motions of residents and turn on and turn off lights automatically and a device to monitor a pulse of a patient and sent data to a medical staff or emergency unit in case of needed. Another example is a system to control a light through web browser [17]. The research aims to create the light controlling system via Wi-Fi using web browser. It can be used in a building, a house, or others. The system also informs the statuses of the lights through web browser. From user satisfaction evaluation for usage of the system, users are very satisfied with the system, 4.19 out of 5.

In automatic system, a microcontroller is the center of data processing both monitoring and controlling. For example, the timing system to switching on and off eight electric appliances are created [18]. A microcontroller, MCS-51, is used to control each relay. The output voltages are at three levels, 12, 24, and 220 v. The timer is set to turn on and off for each output and the status of each are shown on an LCD screen. The result shows that the system is error-free. Another example, NodeMCU which is a microcontroller board with an important Wi-Fi module, ESP8266, to connect to the Internet. The board is similar to Arduino board which comes with input and output ports, programmable to control devices via output ports, and supporting Arduino IDE and C/C++ language. Therefore, the board can be applied for many purposes in IoT environment [19]. NodeMCU is used to control and sending status of a light in the research [20]. It can be applied for another purpose too. ESP8266 and camera module are used for a home security [21]. The notification of the system is done through social media network. There are free services by NETPIE (Cloud Service), Dropbox, Line Notify, and IFTTT that can be used. So, the notification is done in a modern way and real-time to the smartphone of user.

Another important part of IoT system is the interface for users. This can be an application on a smartphone or a website. For example, a mobile application to control animal feeding machine is developed with five functions, feeding animal food, feeding water based on sensor, motion detection of animals, temperature and humidity monitoring, and monitoring by camera module. The system can perform the tasks as expected and the satisfaction of users are very good [22].

Cloud computing is another crucial ingredient for IoT system. It is a service providing computing power, storage, and online application created by service provider. It helps reduce the complexity of installation, maintenance, and time and cost for computing and networking system. They are available in both free and paid services. For example, ThingSpeak is a cloud service for IoT supporting save, share, and visualize data. It uses RESTful Web Service through HTTP protocol and is suitable for Web Apps development. Physical data from sensors can be stored on the system and be displayed as a graph on the web automatically. The system supports import/export data in JSON or XML formats and supports many programming languages, C/C++, Python, Java, JavaScript, etc. [23]. The satisfaction of users to cloud computing [24], they are administrator who are in charge of computer networking system. Four topics are assessed, process, support staff, services on cloud system, and efficiency of cloud system. The result indicates that the satisfaction of users to cloud system is high.

3 Methodology

According to the related works in the previous section, the requirements for the proposed system were collected. Based on those requirements, the system would be designed. Another important part is the study of software development tools, equipment, and tools.

The proposed aquaponics system should be able to monitor temperature and humidity. The light can be turn on automatically during the night time and turn off during the day time. The moisture sensor is used to monitor the level of the water in the hydroponics system. When the water level is low, the pump is started to bring water in from the fish tank to the hydroponics system and let it flow back to the tank repeatedly. So, the plant would use up nitrate in the water from the fish tank, this is water cleaning process. Depending on the type of plant and fish, all related factors need to be adjusted accordingly. The automatic fish feeder would be installed in the system. All of these are controlled and connected to the Arduino microcontroller board and the board connect to a router which connected to the Internet. Arduino Integrated Development Environment (IDE) is used to implement the software for the board. The Google cloud database is used to store data from the system. An IoT platform called Blynk is used to create mobile application (see Fig. 1). The part of hydroponics system is designed using SketchUp software. The structure and layout can be adjusted until meet the requirements before implemented (see Fig. 2).

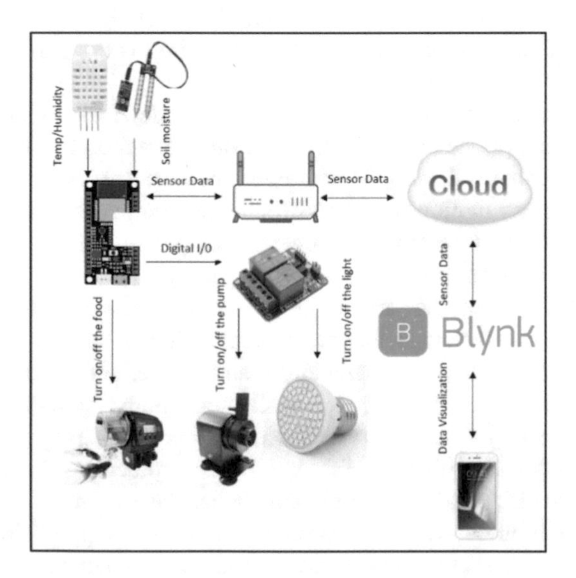

Fig. 1. The proposed aquaponics system.

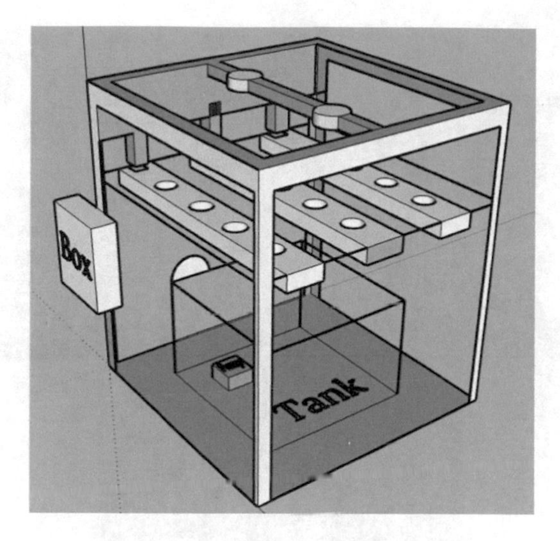

Fig. 2. The part of hydroponics system.

4 Results

The proposed system was created using PVC tube and connectors as a structure of hydroponics part over the fish tank (see Fig. 3). The Arduino microcontroller board is sealed in the box as shown in Fig. 3 on the left to protect it from water and humidity, it is the center of controlling and processing of the system. The software was developed and installed on the chip of this Arduino board. The humidity, temperature, and moisture sensor are installed in the system connected to the Arduino board as shown in Fig. 3 on the right. The fish tank is under the hydroponics system so the water can flow down easily. The pump is in the fish tank and connected to the hydroponics gully. The automatic fish feeder is working on the timer. The plastic cover would help keep the level of humidity in case of needed, otherwise this can be open up. The light is installed at the top of the system to provide the brightness when needed and controlled by the Arduino board.

The plants are tested to live in the aquaponics system (see Fig. 4). The water was flowing from the fish tank to the plants and back cyclically. The data from the sensors are sent to the database on the cloud via the Arduino board Internet connection. The information of humidity, temperature, moisture (water level in hydroponics system), and status of the lights are showing on the dashboard of the mobile application developed by Blynk platform (see Fig. 5). The application can be used to set times for automatic fish feeder and to control the light to be turn on and off manually as well.

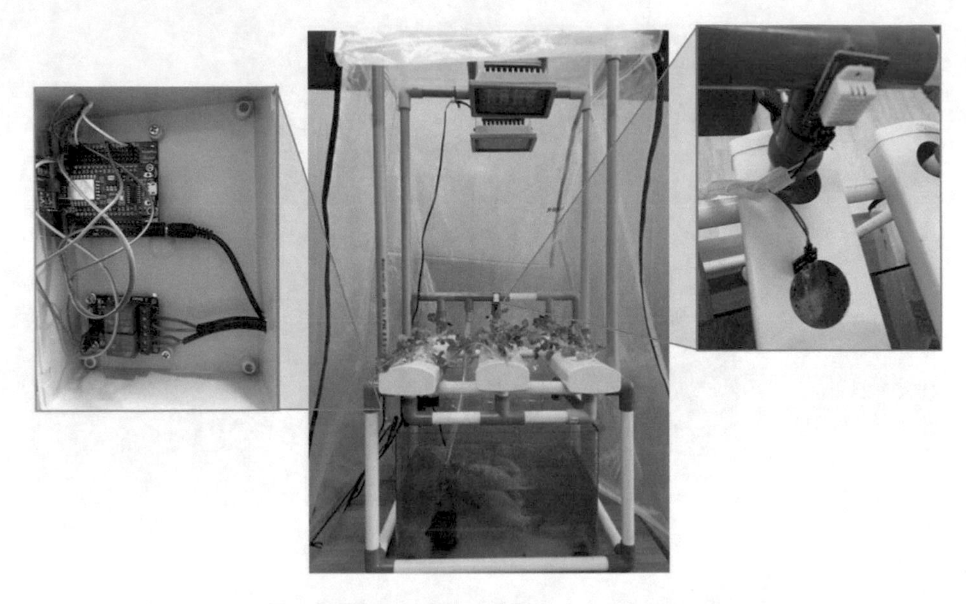

Fig. 3. The structure of the proposed system.

Fig. 4. The plants in the aquaponics system.

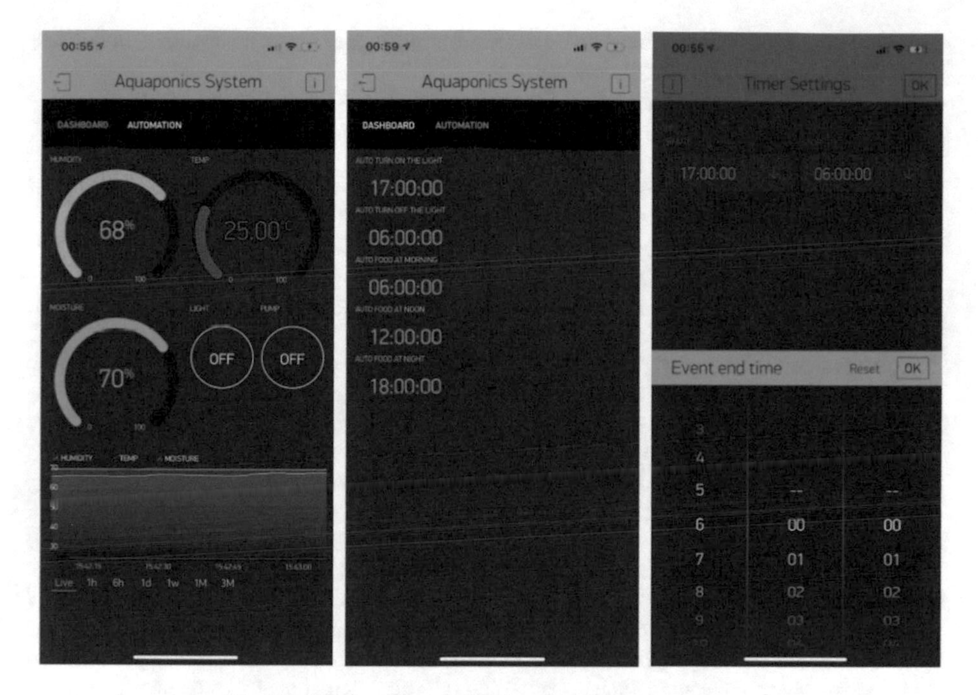

Fig. 5. The mobile application of aquaponics system.

5 Conclusion

The proposed aquaponics system is implemented successfully. Even though the structure is made of plastic, it is strong enough to hold the weight of the hydroponics system. The water cycling system work as expected. The Arduino board received data from sensors and sent them to the cloud via wireless Internet. Therefore, the IoT system is working as designed. The automation part is controlled by the board locally, but the settings can be done through the mobile application. The important information is show on the dashboard of the application and the lights are controllable by clicking on the button of the application. Overall, the system is working well as planned.

The future work is making the system more automatic and be able to control more factors related to the growth of plant and fish, such as humidity, temperature, light level, depends on type of plant. More study needs to be done to find those new factors. Some parts of the system can be tidy up more, such as, the installation of the sensors. The data from the system should be analyzed to find more insightful information.

References

1. Wastewater Treatment by Phytoremediation Methods. https://www.hii.or.th/wiki84/index.php/การบำบัดน้ำเสียด้วยระบบพืชกรองน้ำเสีย. Accessed 20 July 2019
2. Understanding Internet of Things (IoT). http://www.veedvil.com/news/internet-of-things-iot/. Accessed 17 July 2019

3. The Internet of Things. http://its.sut.ac.th/index.php?option=com_content&view=article&id=72&Itemid=46. Access 22 Aug 2018
4. Password security. http://oho.ipst.ac.th/secure-password/. Accessed 22 Aug 2018
5. What is agriculture 4.0? https://www.organicfarmthailand.com/what-is-agriculture-4-0/. Accessed 09 Sept 2019
6. Treehajindarat, T., Punjard, T., Kotcharin, P.: Smart Garden System of IoT. Faculty of Engineering, Srinakharinwirot University (2016)
7. Thongpan, N., Tiengpak, T.: Automatic watering systems via wireless sensor network. J. Inf. Technol. Manag. Innov. **3**(1), 35–43 (2016)
8. Smart Farming Using Drone. https://scitech.kpru.ac.th/portal/content/93CITxVgA24. Accessed 06 Sept 2019
9. Aquaponics, balance of fish and plant. https://www.organicfarmthailand.com/what-is-aquaponics/. Accessed 09 Sept 2019
10. Chowwalittrakul, S.: Automatic Hydroponic Vegetables Growing System. Graduate School, Bangkok University (2018)
11. Aquaponics, new technology. https://www.technologychaoban.com/news-slide/article_90156. Accessed 22 Aug 2019
12. Triyarat, W.: Optimization of Nitrate Treatment System for Marine Shirmp Culture Tank using the Tubular Denitrification Reactor. Chulalongkorn University (2003)
13. Aquaponics Manual. http://www.thai-explore.net/file_upload/submitter/file_doc/3a9f4a45ef87d3bb53dabd9bb1136133.pdf. Accessed 22 Aug 2018
14. Ruengray, P., Wanichpongpan, P., Attasat, S.: The effects of plant ratios on the performance of recirculating aquaponic system. In: National and International Conference Interdisciplinary Research for Local Development sustainability, pp. 99–110. Graduate Studies in Northern Rajabhat Universities, Thailand (2015)
15. Aquaponics – How Many Plants Per Fish? https://www.leaffin.com/aquaponic-fish-plant-ratio/. Accessed 19 July 2019
16. What is Internet of Things. https://blog.sogoodweb.com/Article/Detail/59554. Accessed 19 July 2019
17. Yuenyongphanit, S.: Controlling system to turn on-off the lights through web browser. In: 12th Naresuan Research Conference, pp. 197–203. Naresuan University, Thailand (2016)
18. Eiamsakun, S., Janngoen, S., Kumpomg N., Burana, W., Supabowornsathian, V.: The ON-OFF Timing Control Set for Eight Appliances Using Microcontroller. SIAM University (2013)
19. Beginning IoT with NodeMCU. https://medium.com/@benz20003/iot-ไปกับ-nodemcu-ฉบับเริ่มต้น. Accessed 12 July 2019
20. Kulyot, J., Chavipat, S.: Wireless light switch prototype with an android application. In: 4th Kamphaeng Phet Rajabhat University National Conference, pp. 1388–1393. Kamphaeng Phet Rajabhat University, Thailand (2017)
21. Luangwongsakorn, P.: Home Security by ESP8266. Mahanakorn University of Technology, Thailand (2016)
22. Thaenthong, J., Takaew, S., Sornphrakhanchai, K.: Mobile application development for pet feeder with using microcontroller and Internet of Things. J. Inf. Sci. Technol. **9**(1), 28–40 (2019)
23. Could Computing and Cloud Definition. https://www.it24hrs.com/2015/cloud-computing-and-cloud-definition/. Accessed 09 Sept 2019
24. Chavipat, S., Nualkaew, J., Khotcharith, S. Satisfaction of using cloud computing system of Kamphaeng Phet Rajabhat University. In: 2th Kamphaeng Phet Rajabhat University National Conference, pp. 115–125. Kamphaeng Phet Rajabhat University, Thailand (2015)

Classification of Generation of Thai Facebook Users Using Deep Learning with Probability of Words

Suppachai Tangtreerat(✉) and Sukree Sinthupinyo

Department of Computer Engineering, Chulalongkorn University, Bangkok, Thailand
6070974421@student.chula.ac.th,
sukree@cp.eng.chula.ac.th

Abstract. Facebook is the most popular platform in the world. Marketers would like to use Facebook user data, which comprises large amounts of information which is useful for marketing. Therefore, analyzing the generation of Facebook users for marketing research is important to successfully capture the target market. In this research, posted data of Thai Facebook users will be analyzed using the combined methods of deep learning and probability of words data. The experiment result yields an accuracy of 82.90% per user and 52.48% per status, which is better than using other models alone such as Multi-Layers Perceptron (MLP), Convolution Neural Networks (CNN), or Long Short-Term Memory (LSTM). The experiment results show that using probability of words in each generation can help to increase the efficiency of the model.

Keywords: Generation classification · Thai Facebook · Deep learning · Convolutional neural network · Long-Short Term Memory

1 Introduction

Nowadays, Thai people tend to use Facebook to buy products or advertise services. Generation data is important information in marketing research. However, using only data such as age or date of birth to determine the target market still presents an issue because many users do not fill in age information or fill in fake information.

Machine Learning (ML) is a standard method for text classification. First of all, text representation by vector using Bag-Of-Words or TF-IDF is used, followed by classification by ML algorithms: e.g. Support Vector Machine (SVM), Naive Bayes (NB), Decision Tree, or Multi-Layers Perceptron (MLP). Classifying emotion in Thai YouTube comments [1] is a good example of using this method.

Deep Learning is another method for text classification, which has better accuracy than Machine Learning. Therefore text representation by Word Embedding precedes classification by Deep Learning algorithms: e.g. Convolutional Neural Network (CNN), or Long-Short Term Memory (LSTM).

In [2], a questionnaire was used to collect slang words between Thai teenagers and adults: e.g., "แจ่ม", "แอ๊บ" and "จัดหนัก". Teenagers have 100% usage of the words "แจ่ม" and

P. Meesad and S. Sodsee (Eds.): IC^{2}IT 2020, AISC 1149, pp. 49–59, 2020.
https://doi.org/10.1007/978-3-030-44044-2_6

"แอ๊บ", but only 40% for adults. The word "จัดหนัก" has a 96.6% usage by teenagers and 30% usage by adults. For the results from the questionnaire, we can use the percentage of using words in each group to classify teenagers and adults.

In this work, we would like to compare several word representation methods e.g. Bag-Of-Words, TF-IDF, word2vec and probability of words. We propose the classification of generation of Thai Facebook users using deep learning with probability of words to increase the efficiency of classification of generation.

2 Related Works

2.1 Research of Using Words in Each Age Range

Apisara Pholnarat [2] proposed "Distinctive verbs of Thai teenagers' speech" using a questionnaire about using slang words between two groups: teenagers and adults. The teenager group had an age range from 11 to 20 years old, while the adult group had an age range of 40 and above. Both groups consisted of 50 people. The experiment result showed distinctive usage of slang words such as "มโน" (imagine), "นอย" (feeling bad), and "แซ่บ" (sexy) which were used by teenagers more than adults.

2.2 Research of Classification Age of Social Network Users

Simaki [3] presented the article "Age Identification of Twitter Users: Classification Methods and Sociolinguistic Analysis". The method used in the article is to split each Twitter post into words and create three feature extraction methods. Firstly, text mining features, whereby Twitter posts were translated into 40 dimension vectors. Secondly, sociolinguistic features, whereby theoretical sociolinguistic markers of linguistic differentiation have a length equal to 6. Thirdly, context-based features, whereby some features such as the normalized number of future tense uses, the normalized number of self-references, or the normalized number of hyperlink uses can be used to translate Twitter posts into vectors that have a length equal to 3. Then, a machine learning method is used to classify vectors of the data into 6 specific age ranges: 14–19, 20–24, 25–34, 35–44, 45–59, and 60 or above. The Random Forest method is one of the most accurate methods with an accuracy of 61%.

Bayot [4], the author of "Age and Gender Classification of Tweets Using Convolution Neural Network 2018" used tweet data of users in England and Spain provided in PAN 2016 [5]. In the work, created words corpus by using word2vec [6] by using data from English and Spanish Wikipedia, defining output dimensions of word vectors equal to 100 and 300. Then, he converted the dataset of PAN 2016 to matrix by words corpus. Next, pass the matrix data were passed through the CNN method, proposed by Kim [7]. The data were classified into 5 specific age ranges including, 18–24, 25–34, 35–49, 50–64, and 65 and above, and 2 specific genders: male and female. The experiment results of the CNN method have higher accuracy than SVN classifiers trained on TF-IDF.

2.3 Research of Classification Thai Language Text

Koomsubha [8] presented "Text Categorization for Thai Corpus using Character-Level Convolution Neural Network". The author used news from news agencies in Thailand. The work uses character-level one hot vector text representation. A total of 151 characters were used, including 81 Thai characters, and 70 English and special characters. Then, the matrix data were passed through CNN. This method was more accurate than older CNN methods and machine learning.

Charoenkwan [9] presented "A Sentimental Analysis Using Deep Learning Combined with Bag-of-Words features on Thai Facebook Data". The author created two sets of data with different methods. The first method used Long Short-Term Memory (LSTM) with text representation by word2vec. The second method used Multi-Layers Perceptron (MLP) with text representation by Bag-of-Words. Then, the two methods were com-bined by fully connected layers. This combination method was more accurate than using only LSTM or only MLP.

3 Research Experiments

3.1 Data

Data used in the experiment were collected from Thai Facebook users. A total of 9000 posts were collected from 300 users, 30 posts per user. In this work, data will be categorized into three groups: Generation X (1965–1979), Generation Y (1980–2001), and Generation Z (2002–2020) [10]. Each group contains 100 users.

3.2 Preprocessing

The goal of preprocessing is to transform the posted data of Facebook users to numerical data. Then, the transformed data can be used in machine learning or deep learning. The first step starts with word segmentation. Unlike English, German, or Spanish, the Thai language does not use spaces to separate each word, or full stops (.) to finish sentences. Then, the Deepcut and PyThaiNLP libraries will be used to separate sentences into words. Then, the punctuation and stop words are removed with reference to the Deepcut and PyThaiNLP libraries. In this process, we do not remove emoticons because we want to test whether emoticons affect the classification of generations or not.

3.3 Word Embedding

In this process, we use word embedding, processed by word2vec. We collected 130,000 posts from Thai Twitter users. Then we used the Deepcut and PyThaiNLP libraries for word segmentation. Then, punctuation and stop words were removed, with reference to the Deepcut and PyThaiNLP libraries. After processing word2vec successfully, that created a Thai corpus containing 147,611 words.

3.4 Probability of Words

In this process, Probability of words is used in each of the generations for increasing the efficiency of classification generation. Probability calculations are as follows:

$$aw = wc1 + wc2 + wc3 \quad (1)$$

$$vector[word] = [wc1/aw, wc2/aw, wc3/aw] \quad (2)$$

wc1 denotes the number of one word found in generation X, wc2 the denotes the number of one word found in generation Y, wc3 denotes the number of one word found in generation Z, aw denotes the sum of one word.

3.5 Convolution Neural Network

Convolution Neural Network (CNN) was originally designed for use in character image recognition research [11]. Later, CNN was also applied to text analysis. The structure of the Convolution Neural Network included the Convolutional Layer, Pooling Layer and Fully Connected Layer. In the research of KIM [7], a combination of CNN and word2vec was used for sentimental analysis. In the research of Roy Khristopher Bayot [4], CNN and word2vec were used for age and gender classification. In this research, the CNN model of KIM [7] will be used for classifying the generation of Thai Facebook users with additional use of BOW, TF-IDF, and probability of words for improving the accuracy of the result.

3.6 Long-Short Term Memory (LSTM)

Long short term memory (LSTM) [12] is one type of recurrent neural network (RNN) that was designed to solve vanishing gradient problems of recurrent neural networks. If output of Forget Gate is equal to 0 then it removes the previous cell state, but if Forget Gate is equal to 1 it keeps the cell state. Input Gates are responsible for decisions to improve the value or not. Output Gates are responsible for control of the value that goes forward to the next step.

4 Methodology

In this research, focusing on the classification of generations of Thai Facebook users, we will be categorizing the data into 3 groups, namely Generation X, Generation Y, and Generation Z. Then, we will use deep learning combined with Probability of Words in each generation.

We collected 130,000 posts from Thai Twitter users then removed the stop words and punctuation symbols. Then, the Twitter data were processed by word2vec to create a Thai corpus for further usage. After that, the collected data from 300 Facebook users at 30 posts per user, do preprocessing then convert preprocessing data from Facebook to matrix by two methods are as follows: 1. Converting text using probability of words in each generation. 2. Converting text using the Thai corpus from Twitter data. Then, the data forms were passed through to deep learning.

In this research, a 10-fold cross-validation method is used. The dataset was split into two parts: the first part of 90% was then spilt to 80% (6,480 posts) to be used for training and 20% (1,620 posts) for validation. The second part of 10% (900 posts) was to be used for testing.

Accuracy measurement is divided into two methods. The first method is accuracy per status, that uses the results from the model in the format [%Gen X, %Gen Y, %Gen Z] then selects the generation from the greatest percentage. The second method is accuracy per user, that uses the sum of the results from the model by grouping users then selecting the generation from the greatest percentage.

5 Experiment Result

The experiment of this research uses the Keras framework of python, a system run on Google Colab, and uses GPU runtime. In this experiment three models are tested, namely MLP, CNN, and LSTM to compare the efficiency of each model.

We created the MLP model, defining the output to have three labels, X,Y, and Z, and used two text representation methods, namely Bag-of-Words and TF-IDF, and used a grid search to find the best parameters. The grid search defined the parameters as follows: max_iter = [200], hidden_layer_sizes = [50, 100, [150, 50]], learning_rate_init = [0.001, 0.0001].

Table 1. Prediction accuracy (%) of MLP.

Model	Tokenize	ACC (%)	
		User	Status
MLP (BOW)	DEEP CUT	77.30	50.90
MLP (TF-IDF)	DEEP CUT	80.00	51.25

Table 2. Precision, recall, F1 (%) of MLP.

Model	Label	User			Status		
		Precision	Recall	F1	Precision	Recall	F1
MLP (BOW)	GEN X	76.86	72.00	74.03	53.25	44.13	48.23
	GEN Y	72.83	63.00	66.59	45.22	38.43	41.53
	GEN Z	83.64	97.00	89.40	53.20	70.13	60.47
MLP (TF-IDF)	GEN X	77.87	71.00	73.32	51.78	46.33	48.86
	GEN Y	71.64	74.00	72.36	44.13	42.53	43.24
	GEN Z	92.84	95.00	93.59	57.07	64.90	60.62

Tables 1 and 2 show the experiment accuracy of the MLP model. The best of the results from the MLP model uses the Deepcut library for tokenization sentences and uses text representation by TF-IDF, that gets the best accuracy of test set, accuracy per user equal to 80.00%, and accuracy per status equal to 51.25%. F1 per user of Gen X was equal to 73.32%, Gen Y was equal to 72.36%, and Gen Z was equal to 93.59%. F1 per status of Gen X was equal to 48.86%, Gen Y was equal to 43.24%, and Gen Z was equal to 60.62%.

We created the CNN and LSTM models by using four text representation methods, namely word2vec, word2vec and Bag-of-Words, word2vec and TF-IDF, and word2vec and Probability of Words. In the training models of CNN and LSTM, the defined optimizer algorithm is Adaptive Moment Estimation (Adam), the defined loss function is Binary Cross-entropy, the defined batch size was equal to 50, and the four defined output types are as follows: 1. Gen X, Y, Z; 2. Gen X, Y; 3. Gen X, Z; 4. Gen Y, Z.

Table 3. Prediction accuracy (%) of CNN, LSTM (3 labels).

Model	ACC (%)	
	User	Status
CNN (word2vec)	74.33	48.36
CNN (word2vec + BOW)	78.00	51.04
CNN (word2vec + TF-IDF)	81.33	50.75
CNN (word2vec + Prob of words)	**82.90**	**52.48**
LSTM (word2vec)	69.33	47.74
LSTM (word2vec + BOW)	78.00	50.80
LSTM (word2vec + TF-IDF)	78.66	51.24
LSTM (word2vec + Prob of words)	81.66	51.84

Table 4. Precision, recall, F1 (%) of CNN, LSTM (3 labels).

Model	Label	User			Status		
		Precision	Recall	F1	Precision	Recall	F1
CNN (word2vec + Prop of words)	GEN X	79.43	82.00	79.89	53.65	49.76	51.49
	GEN Y	80.09	72.00	74.40	45.58	42.50	43.80
	GEN Z	92.75	95.00	93.59	57.66	65.20	61.07

Tables 3 and 4 show the experiment accuracy of the CNN and LSTM models (three labels). The best of the results from these models is the CNN model using the Deepcut library for tokenization sentences and using text representation by word2vec combined with Probability of Words, that provides the best of accuracy of the test set, accuracy per user equal to 82.90%, and accuracy per status equal to 52.48%. F1 per user of Gen X was equal to 79.89%, Gen Y was equal to 74.40%, and Gen Z was equal to

93.59%. F1 per status of Gen X was equal to 51.49%, Gen Y was equal to 43.80%, and Gen Z was equal to 61.07%.

Table 5. Prediction accuracy (%) of CNN, LSTM (2 labels).

Model	ACC (%) per user			ACC (%) per status		
	X, Y	X, Z	Y, Z	X, Y	X, Z	Y, Z
CNN (word2vec)	73.50	95.00	82.50	58.58	71.01	63.56
CNN (word2vec + BOW)	76.50	97.00	90.00	58.26	73.14	66.45
CNN (word2vec + TF-IDF)	75.50	97.00	92.50	58.33	73.75	67.00
CNN (word2vec + Prob of words)	**80.00**	**98.00**	**93.00**	**58.35**	**73.86**	**67.01**
LSTM (word2vec)	74.00	95.00	85.50	58.20	71.16	63.85
LSTM (word2vec + BOW)	79.49	96.49	90.50	58.28	73.16	65.09
LSTM (word2vec + TF-IDF)	77.00	97.00	92.50	58.70	73.10	66.28
LSTM (word2vec + Prob of words)	**79.00**	**95.50**	**94.00**	**58.00**	**72.04**	**67.73**

Table 6. Precision, recall, F1 (%) of CNN, LSTM (2 labels).

Model	Label	User			Status		
		Precision	Recall	F1	Precision	Recall	F1
CNN (word2vec + Prop of words)	GEN X	83.35	76.00	78.79	58.55	57.10	57.80
	GEN Y	78.70	84.00	80.85	58.17	59.60	58.86
CNN (word2vec + Prop of words)	GEN X	96.51	100.00	98.13	74.29	73.33	73.70
	GEN Z	100.00	96.00	97.83	73.69	74.40	73.95
CNN (word2vec + Prop of words)	GEN Y	90.10	97.00	93.26	66.79	67.90	67.28
	GEN Z	97.07	89.00	92.66	67.38	66.13	66.68
LSTM (word2vec + Prop of words)	GEN X	76.56	87.00	80.53	56.58	68.86	62.05
	GEN Y	86.41	71.00	76.58	60.37	47.13	52.79
LSTM (word2vec + Prop of words)	GEN X	92.59	100.00	95.95	69.84	77.60	73.47
	GEN Z	100.00	91.00	94.86	74.97	66.50	70.41
LSTM (word2vec + Prop of words)	GEN Y	92.34	96.00	94.04	70.11	62.00	65.74
	GEN Z	96.18	92.00	93.93	65.94	73.46	69.46

Tables 5 and 6 show the experiment accuracy of the CNN and LSTM models (2 labels). The best of the results from this model is the CNN model, using text representation by word2vec combined with probability of words, which provides the best accuracy of test set, accuracy per user of X, Y label equal to 80.00%, X, Z label equal to 98.00%, Y, Z label equal to 93.00%, and accuracy per status of X, Y label equal to 58.35%, X, Z label equal to 73.86%, and Y, Z label equal to 67.01%. The second one is the LSTM mod-el using text representation by word2vec combined with probability of word, with accuracy per user of X, Y label equal to 79.00%, X, Z label equal to 95.50%, and Y, Z label equal to 94.00%, and accuracy per status of X, Y label equal to 58.00%, X, Z label equal to 72.04%, and Y, Z label equal to 67.73%.

The experiment of CNN and LSTM, when we used word2vec combined with BOW or TF-IDF or Probability of Words and trained on the CNN and LSTM models, had better efficiency than when using the CNN or LSTM models with text representation by word2vec alone.

Table 7. Confusion matrix table for per user, per status.

		Predicted class		
		X	Y	Z
Actual class (per usre)	X	82	18	0
	Y	20	72	8
	Z	3	2	95
Actual class (per usre)	X	1,493	910	597
	Y	872	1,275	853
	Z	428	616	1,956

Table 7 show the sum confusion matrix from 10-fold cross-validation of the CNN model using text representation by word2vec combined with probability of word of the test set. Table 7 shows results per user, where the test set had 300 users and each generation had 100 users. TP of Gen X was equal to 82, TP of Gen Y was equal to 72, and TP of Gen Z was equal to 95. Table 8 shows results per status, where the test set had 9000 statuses and each generation had 3000 statuses. TP of Gen X was equal to 1,493, TP of Gen Y was equal to 1,275, and TP of Gen Z was equal to 1,956.

The accuracy table and confusion matrix table can be summarized as follows: (1) Result per user can increase efficiency of prediction of generation because any single status of user does not indicate the generation but when we sum the result by grouping users then select the generation from the greatest percentage this can increase efficiency. (2) In this experiment, classification of generation has 2 labels, that include Gen X, Z for the best classification, and Gen Y, Z respectively. (3) In this experiment, the best classification is Gen Z, Gen X, and Gen Y respectively. (4) Word segmentation has an effect on the efficiency of the model. The experiment uses the Deepcut library which has greater efficiency than the PyThaiNLP library.

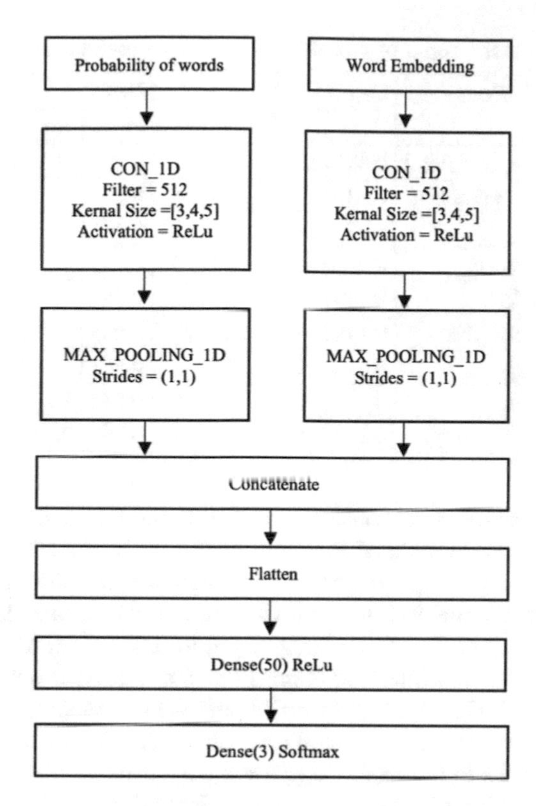

Fig. 1. Architecture of CNN (word2vec + probability of words).

Figure 1 shows the structure of the CNN model using text representation by word2vec combined with Probability of Words. After already completing the text representation by word2vec and text representation by Probability of Words, we sent the data to 1D convolution layer, define filter = 512, kernel size = [3–5] and activation = ReLu, then sent the result to 1D max pooling layer, define strides = (1, 1), then combined the result from the max pooling layer of word2vec and Probability of Words by using the concatenate layer and flatten layer, and then sent the result to the fully connected layer, define Dense = 50 and Activation = ReLu, and in the last layer, define Dense = 3 and Activation = Softmax.

Table 8. Top 3 of probability of words in each generation.

Word	Gen X		Gen Y		Gen Z	
	Prob.	No.	Prob.	No.	Prob.	No.
สถานที่ (place)	0.857	12	0.142	2	0	0
ครัว (kitchen)	0.83	20	0.166	4	0	0
😝	0.77	57	0.108	8	0.121	9
🌵	0	0	0.857	12	0.142	2
🍭	0.166	4	0.791	19	0.041	1
ปลา (fish)	0.178	5	0.785	22	0.035	1
🔥	0	0	0.138	5	0.861	31
ป่ะ	0	0	0.157	3	0.842	16
ค่าย (camp)	0.172	5	0.034	1	0.793	23

Table 8 shows the top 3 of Probability of Words and the number of words found in each generation. From the data of Probability of Words in each generation, we can classify generations, e.g. "สถานที่" (place) has the highest Probability of Words in Gen X equal to 0.857, Gen Y equal to 0.142, but was not found in Gen Z. Emoticon "🔥" has the highest Probability of Words in Gen Z equal to 0.861, Gen Y equal to 0.138, but was not found in Gen X, while "ค่าย" (camp) has the highest Probability of Words in Gen Z equal to 0.793, Gen X equal to 0.172, and Gen Y equal to 0.034. Using Probability of Words to help classify generations can create a rule base by which humans can predict generations to the model, e.g. users posted status including the word "ค่าย" (camp), they have a high probability to be a student which is Gen Z. Or users posted status about the word "ครัว" (kitchen), they have a high probability to be an adult, which are Gen X and Gen Y.

6 Conclusion

In this research, we propose the classification of generations of Thai Facebook users. We use data from Thai Facebook users and use text representation by word2vec, that is processed by using data from Twitter. This research added a new feature as Probability of Words to help improve the efficiency of the model and then processed data by the CNN model for the classification of generations of Facebook users. The results showed an accuracy per user equal to 82.90% and accuracy per status equal to 52.48%, which was better accuracy than when using word2vec alone by about 8%. Furthermore, Probability of Words can help to classify generations and can create a rule base whereby humans can predict generations: e.g. if users posted about the word "สอบ" (exam), they have a high probability to be a student.

In future works, we will collect more Thai Twitter data to create a larger Thai corpus and collect more Thai Facebook users to improve the accuracy of the model.

References

1. Sarakit, P., Theeramunkong, T., Haruechaiyasak, C., Okumura, M.: Classifying emotion in Thai Youtube comments. In: 2015 6th International Conference of Information and Communication Technology for Embedded Systems (IC-ICTES), Hua-Hin, pp. 1–5 (2015)
2. Pholnarat, A.: Distinctive verbs of Thai teenagers' speech. J. Lang. Cult. **35**(Special), 231–235 (2016)
3. Simaki, V., Mporas, I., Megalooikonomou, V.: Age identification of Twitter users: classification methods and sociolinguistic analysis. In: Gelbukh, A. (ed.) Computational Linguistics and Intelligent Text Processing, CICLing 2016. LNCS, vol. 9624. Springer, Cham (2018)
4. Bayot, R.K., Gonçalves, T.: Age and gender classification of tweets using convolutional neural networks. In: Nicosia, G., Pardalos, P., Giuffrida, G., Umeton, R. (eds.) Machine Learning, Optimization, and Big Data, MOD 2017. LNCS, vol. 10710. Springer, Cham (2018)
5. Rangel, F., Rosso, P., Verhoeven, B., Daelemans, W., Pottast, M., Stein, B.: Overview of the 4th author profiling task at PAN 2016: cross-genre evaluations. In: Balog, K., Cappellato, L., Ferro, N., Macdonald, C. (eds.) Working Notes Papers of the CLEF 2015 Evaluation Labs. CEUR Workshop Proceedings, vol. 1609, pp. 750–784 (2016)
6. Mikolov, T., Sutskever, I., Chen, K., Corrado, G., Dean, J.: Distributed representations of words and phrases and their compositionality. In: Burges, C., Bottou, L., Welling, M., Ghahramani, Z., Weinberger, K. (eds.) Advances in Neural Information Processing Systems, vol. 26, pp. 3111–3119 (2013)
7. Kim, Y.: Convolutional Neural Networks for Sentence Classification. CoRR abs/1408.5882 (2014)
8. Koomsubha, T., Vateekul, P.: A character-level convolutional neural network with dynamic input length for Thai text categorization. In: 2017 9th International Conference on Knowledge and Smart Technology (KST), pp. 101–105. IEEE, Chonburi (2017)
9. Charoenkwan, P.: ThaiFBDeep: a sentimental analysis using deep learning combined with bag-of-words features on Thai Facebook data. In: 2018 7th International Congress on Advanced Applied Informatics (IIAI-AAI), pp. 565–569. IEEE, Yonago (2018)
10. Berkup, S.B.: Working with generations X and Y in generation Z period: management of different generations in business life. Mediterr. J. Soc. Sci. **5**(19), 218 (2014)
11. Lecun, Y., Bottou, L., Bengio, Y., Haffner, P.: Gradient-based learning applied to document recognition. Proc. IEEE **86**(11), 2278–2324 (1998)
12. Hochreiter, S., Schmidhuber, J.: Long short-term memory. Neural Comput. **9**(8), 1735–1780 (1997)

Design of an Intelligent, Safe and Secure Transport Unit for the Physical Internet

Roman Gumzej[1], Maytiyanin Komkhao[2](✉), and Sunantha Sodsee[3]

[1] Faculty of Logistics, University of Maribor, Celje, Slovenia
roman.gumzej@um.si
[2] Faculty of Science and Technology, Rajamangala University of Technology Phra Nakhon, Bangkok, Thailand
maytiyanin.k@rmutp.ac.th
[3] Faculty of Information Technology, King Mongkut's University of Technology North Bangkok, Bangkok, Thailand
sunantha.s@it.kmutnb.ac.th

Abstract. The number of air shipments with high added value is rapidly growing. For this kind of cargo, it is often necessary to monitor its specific characteristics also during transit. To enable this feature, a special kind of freight containers appropriate for air cargo ought to be employed, which can maintain prescribed micro-climatic conditions inside of them as well as communicate their parameters to their consignors. At the same time, a cooperative intelligent transport system should ensure valuable cargo's protection against loss and damage throughout the entire transport chain. In fulfilment of these requirements an appropriate, safe and secure intelligent transport unit (iTU) is designed meeting the International Air Transport Association's (IATA) standards for transporting pharmaceuticals as well as the principles of the Internet of Things and the arising Physical Internet. Deploying the iTU is expected to render handling and transportation of sensitive, high-value cargo by Cooperative Intelligent Transport Systems (C-ITS) more transparent, safe and efficient.

Keywords: Air transport · Cooperative Intelligent Transport System · Intelligent transport unit · Safety · Security

1 Introduction

The initiatives "Industrie 4.0" (Germany) and "Made in China 2025" (China) stand for smart production of smart products. Enabling the realizations of these concepts, the Intelligent Web (Web 2.0) in general and the Internet of Things (IoT) in particular are needed as infrastructural basis. In a broad sense, anything "smart" needs to be implemented as a Cyber-Physical System (CPS). The Physical Internet (PI) as a reconfigurable transportation network accommodates roaming intelligent Intermodal Loading Units (ILUs), which are characterized by their ability to autonomously navigate their way through the PI. Within the PI, PI hubs are logistic centers with capacities to store and forward ILUs in the form of smart containers. Interlogistics 4.0

P. Meesad and S. Sodsee (Eds.): IC^{2}IT 2020, AISC 1149, pp. 60–69, 2020.
https://doi.org/10.1007/978-3-030-44044-2_7

is operated by Cooperative Intelligent Transport Systems (C-ITS) managing ILUs and moving them throughout the PI like data packets in the Internet. As for the PI, the main criterion for a successful implementation of Interlogistics 4.0 is Quality of Service (QoS).

The number of air shipments with high added value has grown substantially during the past years. High value-to-weight manufactured products such as micro-electronics, pharmaceuticals, live tissue, medical devices and aerospace components are increasingly being transported as air cargo. For contemporary customers as well as Industrie 4.0 production sites the just-in-time principle is not sufficient anymore. In Interlogistics 4.0 greater transparency of the logistic processes is required, as the customers do not only demand to know at any time their sensitive cargo's current location and estimated arrival time, but also to monitor the cargo's complete parameter status during transportation. To address such a complex scenario, cargo operators must employ C-ITS at their Interlogistics 4.0 hubs as well as utilize intelligent transport units designed to enable appropriate monitoring and routing of valuable cargo throughout the PI.

This article focuses on the construction of an intelligent, safe and secure PI Transport Unit called iTU in abbreviation. First, the pertaining IATA regulations for air cargo are reviewed, followed by the relevant C-ITS concepts. Then, the design of the proposed iTU is elaborated from the physical, informational as well as procedural perspectives, and evaluated according to the previously mentioned regulations. In conclusion, the role iTU-based transport automation may play in the framework of C-ITS is discussed.

2 Overview of the Literature

2.1 IATA Strategy and Regulations

According to the IATA Cargo Strategy [1], the ten industry key priorities by 2020 are:

1. enhancing safety,
2. improving security,
3. pushing for smarter regulations,
4. strengthening the value proposition of air cargo,
5. driving efficiency through global standards,
6. modernizing air cargo,
7. improving quality,
8. protecting cash,
9. strengthening partnerships, and
10. building sustainability.

Although these priorities pertain to air transport, that typically represents less than 1% of world trade by volume, but 35% by value, with more higher-value shipments and their increased demand for security they should pertain to most of the world's trade to build sustainability.

Safety remains the first priority of the IATA Cargo Strategy. Some commodities may endanger the safety of carriers, their passengers and/or crews, if not shipped in accordance with stipulated regulations. In this respect growing attention is also directed towards design and use of fire-resistant Unit Load Devices (ULDs), which are used to maintain cargo in specified temperature ranges as well. Being equally critical, security measures need to be both efficient and effective. To be effective, the operation of shippers needs to be greatly transparent, reliable and predictable. To be efficient, compliance with regulations needs to be managed by industry itself in order not to slow down transit times substantially. To drive efficiency, global standards are needed, e.g. to ensure 48 h end-to-end shipping time if demanded by the customer. The vision is to have a paperless industry, and to be able to rely on high-quality data available on demand to all relevant stakeholders. To improve quality, the transportation industry needs to maintain reliability and consistency of its services.

As part of IATA's special cargo handling and temperature control regulations, a new approach for transported pharmaceutical products was presented using logistic cold chains and Interlogistics 4.0 concepts to obtain an even better solution than envisioned by the IATA Cargo Strategy. The IATA special cargo regulations mostly relate to live animals, perishables and pharmaceuticals [2]. Besides dangerous goods, also other special cargo requires regulations and standards for documentation, handling and personnel training.

Transporting healthcare products by air requires complex logistic processes, specific equipment, storage facilities and harmonized handling procedures to maintain shipment integrity. The Temperature Control Regulations (TCR) [2] are a comprehensive guideline, designed to enable stakeholders involved in transport and handling of pharmaceutical products to safely meet the requirements. TCR contains the requirements and standards for transportation and handling of time- and temperature-sensitive healthcare products, including pharmaceutical product information based on WHO guidelines. Moreover, TCR provides access to the most current and efficient practices for pharmaceutical operations by guaranteeing shipments' compliance with international and/or local regulations. TCR includes:

- up-to-date airline and government requirements pertaining to the transport of healthcare and pharmaceutical products,
- requirements for handling, marking and labelling,
- necessary packaging requirements,
- information on handling procedures, and
- on documentation needed when transporting healthcare products.

The industry often requires pre-set environmental parameters to sustain inside intermodal transport units throughout the entire transport chain. Inside sensitive cargo units, environmental parameters must be maintained at constant levels when transporting sensitive goods. Upon arrival at a distribution point, any unit should, according to the protocols prescribed, additionally be checked for tampering by examining its packaging and the seals on its openings.

2.2 C-ITS Concepts

The main components of a generic Conceptual Information Model (CIM) for integrated freight and fleet management are the following:

Physical Internet (PI) – an open global logistic system based on physical, digital and operational interconnectivity as achieved by encapsulation, interfaces and protocols [5]. The Physical Internet does not manipulate physical goods directly, but exclusively containers (*PI containers*) explicitly designed for the Physical Internet and encapsulating physical goods inside of them [5]. It enables efficient, sustainable, adaptive and resilient Interlogistics 4.0 solutions [6].

Internet of Things (IoT) – semantically, IoT [7] represents the worldwide network of interconnected and uniquely addressable objects and is based on standard communication protocols. The novel paradigm is rapidly gaining ground in the scenario of wireless telecommunications. Its basic idea is the pervasive presence of a variety of things or objects such as RFID tags, sensors, actuators or vehicles around us, which are able, by unique addressing schemes, to interact with each other and to cooperate with their neighbors in order to reach common goals [8, 9].

Cyber-physical Systems (CPS) – connect the real (physical) world of objects and things with the virtual (cyber) world of software and services by means of sensors, actuators and embedded computing devices. Such systems are required to consider real-world effects and contexts of informational processes [10]. CPSs may possibly be improved by engineering self-adaptive processes based on the principle of so-called MAPE-K feedback loops, which consist of the phases Monitor (M), Analyze (A), Plan (P) and Execute (E), and which are repeatedly being executed on a knowledge base (K). Here, the concept of MAPE-K feedback loops, as known from engineering of self-adaptive systems [11], is transferred to process execution.

PI Protocols – a cargo information model includes generation and management of digital representations of the physical and functional characteristics of cargo [12, 13]. The result of this modelling process leads to a shared knowledge resource which supports tracking and tracing of cargo from an original place to conveyance, followed by its operational life before arriving at its destination. The resource should be used during the cargo's operational cycle – from deployment until the arrival at the destination. It provides cargo tracking information, such as physical and chemical characteristics, geometry, shape, transportation state, ownership, location etc.

Ontologies – are a formal way to define the structure of knowledge for various domains with nouns representing classes of objects and verbs representing relations between objects. Ontologies resemble class hierarchies in object-oriented programming. As they are meant to represent information on the Internet coming from all sorts of heterogeneous data sources, they are expected to be rather flexible and evolving almost permanently. The Web Ontology Language (OWL) is a family of knowledge representation languages for authoring ontologies characterized by formal semantics. It is built upon the W3C XML standards for objects called Resource Description Framework (RDF). Both, OWL and RDF are to be used to construct and maintain the aforementioned knowledge base (K).

Software Agents – are objects performing autonomous actions to accomplish specific tasks and are meant to perform the monitoring (M), analysis (A), planning (P) and execution (E) phases of MAPE-K feedback loops of the CPSs within the Interlogistics 4.0 network. In the course of this network's implementation, model-based reflex agents (cp. [14] for a classification of agent types) should be employed, since agents cannot perceive their complete environment within a network. Initially, model-based reflex agents select their actions according to movement-related condition-action rules, which only depend on a model of the world, but not on the current perception of the environment. The model-based reflex approach is natural for cargo tracking and tracing, because multiple threads of control naturally match the distributed and ever-changing nature of the underlying data sources affecting higher-level decision-making processes. The approach allows to more easily manage detection of and response to important time-critical information, that may occur at any time from any of a large number of different sources. Hence, model-based reflex agents are considered highly beneficial for clients, customers and various cargo agents from the point of view of keeping track of cargo from deployment to delivery. Besides, they can also assist in recognizing deviations from planned schedules and taking corrective actions by the C-ITS.

With this in mind, in the sequel an original approach is presented, which secures transport chains by providing them with appropriate informational support to enable automated authentication and authorization of consignors and their shipments, and which guarantees undisrupted flow of information along secure transport chains. At the same time, it introduces mechanisms to protect both the transport units and their data while in transit.

3 Design of the iTU

3.1 Physical Structure

From a physical perspective, an iTU must be easy to handle, store, transport, seal, snap to a structure, interlock, load, unload, build and recycle (green logistics) – just like all PI containers [6]. Logistic hubs, e.g. warehouses, utilize fixed and mobile RFID transceivers to monitor passing cargo. A GPS receiver, installed in a monitoring station, can delegate position information to the stored or conveyed PI containers. This way, mobile PI containers can maintain accurate position data even in closed spaces such as warehouses.

The iTU's design comprises two main parts, viz. a container for the proper cargo and an embedded (computer) system charged with monitoring, control and data handling. This controller basically consists of a microcontroller, an RFID interface, a Wi-Fi interface, various sensors for, e.g., temperature, humidity, pressure, acceleration or orientation, and tamper switches. The sensors within the unit and attached to the controller allow for Remote Condition Monitoring, which represents one of the most important services having a fundamentally positive impact on the quality of service of handling PI containers. The ability to access cargo status information in real time is essential for support services, especially since it enables more efficient root cause analysis and solution development. To ensure low power consumption, a positive RFID identification will turn on Wi-Fi radio for communication, which is automatically turned off again as soon as suitable communication protocols have been worked off. A host of measures is taken to effectively protect the iTU and its controller both mechanically as well as against hacking attacks and malware intrusion.

3.2 Information Model

The conceptual information model (cp. Fig. 1) for the iTU corresponds with the European INTACT guidelines on integrated telematics for advanced communication in freight transport.

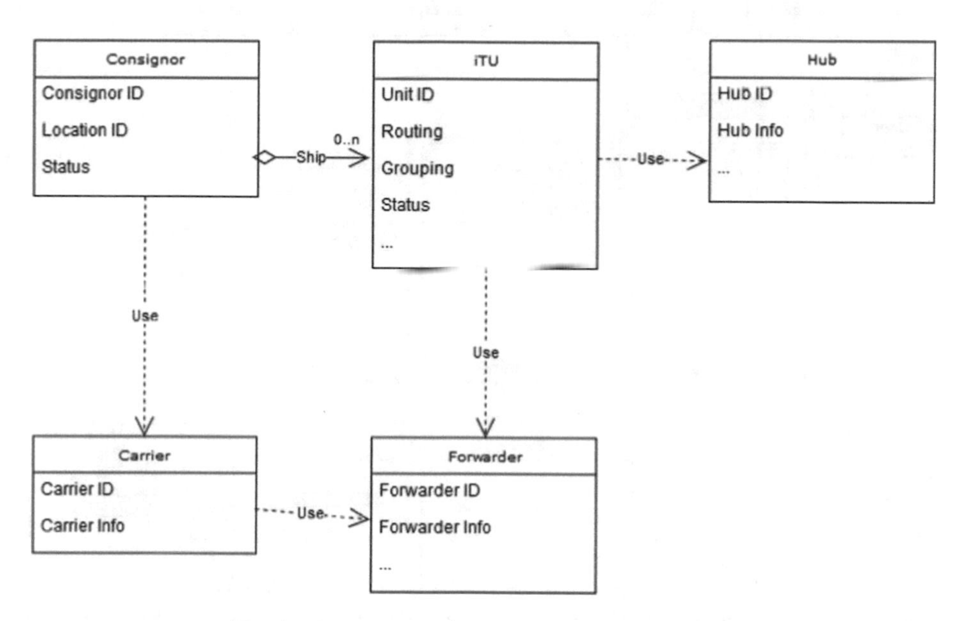

Fig. 1. Conceptual information model for the iTU.

Cargo-related information can be divided into several categorics according to different standards. The first division separates static and dynamic information. Static information about cargo includes cargo identification, date, type, quantity, value of cargo and owner, which are stored in the cargo database. Dynamic information about cargo includes cargo temperature, humidity, routing, status etc., which can initially be stored in the iTU's controller and the cargo database for comparison. If the iTU detects value changes of dynamic information items by comparing current read-outs with the status saved last, it communicates the changes to a back-end C-ITS system. Besides the data mentioned, the cargo database also holds indirect cargo information, including data on order, cargo owner, transportation company and repository [3, 17].

3.3 Functional Model

The Intelligent Agents (IA) technology can be used to handle transportation networks in their entirety. In the process of cargo tracking, all stakeholders as well as the cargo units in a cargo conveyance themselves can be represented by IAs ("digital twins"). If a problem occurs during a conveyance process, the agents can negotiate a plan for problem solving. All agents cooperating in a cargo conveyance system constitute a C-ITS network within the PI. Model-based reflex agents are applied to solve problems that are difficult or impossible to handle by any individual agent alone.

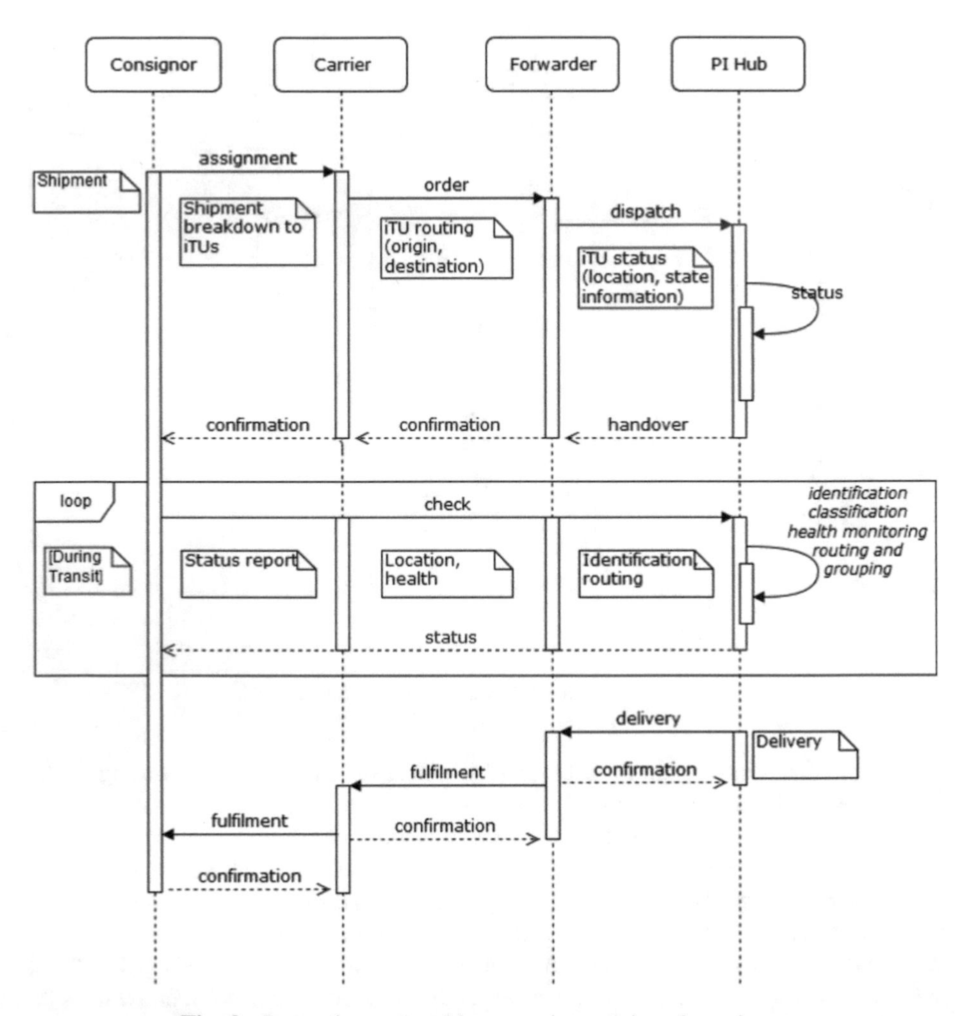

Fig. 2. Interaction protocol between cargo unit and consignor.

Figure 2 presents the collaboration among participants within a C-ITS network in a typical scenario of iTU routing from shipment to delivery. Interaction protocols are defined according to the Consignors', Carriers', Forwarders' and logistic Hubs' standard procedures.

A large shipment may first be broken down to a number of iTUs. They are assigned their origin and destination. Finally, before they are dispatched, their status information is initialized (location and cargo state determined by sensors).

After an iTU is dispatched, it begins its journey towards its destination – from one logistic hub to another in a fastest and most efficient manner. At every logistic hub an iTU's current status and location are automatically logged and forwarded to its forwarder, carrier and consignor. Upon request its current status is also returned to an interested party.

The cargo conveyance procedures are enclosed by a MAPE-K condition feedback loop. When the protocols are being performed, all pertaining participants in a transport chain are acquainted with an iTU's status and can act accordingly, e.g. in case a monitored variable is above or below a given threshold. If all information obtained from the protocol sequences, viz. identification, classification, health monitoring, routing and grouping, yield values corresponding with prescribed rules and thresholds, no action is needed. Otherwise, corresponding actions are requested by appropriate signals through the cargo handling agents to the back-end C-ITS systems.

Upon reaching its final destination, an iTU announces its arrival to its forwarder, carrier and consignor. Once all iTUs of a shipment have arrived, the consignor confirms the fulfillment of the shipping assignment, and the carrier confirms the fulfillment of the shipping orders.

3.4 Safety and Security

Owing to legal ordinances and the increasing risk potential, transport units must be examined for explosives, especially before loading them to airplanes or ships. This process, which may even be needed multiple times throughout a single transport chain, introduces additional overhead causing increased transport costs and slowing down logistic processes. To prevent multiple examinations of transport units after their initial loading, Authorized Economic Operators, Known Consignors as well as Regulated Agents (Directives (EC) No. 648/2005 [18] and (EU) No. 185/2010 [4]), representing producers, distributors and transport hubs, were assigned special ordinances they must obey in order to preserve their status and uphold the required safety and security of their shipments. A protocol [15, 16] for automated authentication and authorization of transport units and their consignors was developed and patented with the objective to speed up logistics processes and to reduce costs.

4 Discussion

Cooperative Intelligent Transport Systems (C-ITS) are considered ubiquitous as part of Interlogistics 4.0. The purpose of C-ITS is to make transport more efficient, clean, safe and cost-effective as well as to offer extensive opportunities for the development of business and innovation ideas in the transport domain. The next big step in their introduction was to construct the PI, along which iTUs can be transported.

As part of the envisaged transportation processes, also the new concept of Structural Health Monitoring [19] for handling environmental and operational damages on smart shipping containers should be adopted with the final goal of autonomous process adaptation within C-ITSs to deal with different extraordinary scenarios. By this approach, the unit's autonomy and resilience would increase throughout conveyance processes due to self-adaptation and self-healing capabilities. Employing process meta-models and execution routines in implementing feedback loops, the proposed solution would eventually render a completely autonomous PI.

5 Conclusion

To be able to handle air shipments with high added value by C-ITS, a special kind of freight containers appropriate for air cargo was designed, which can maintain prescribed micro-climatic conditions inside of them as well as communicate their parameters to their consignors. Our safe and secure intelligent transport unit (iTU) is designed meeting the International Air Transport Association's (IATA) standards for transporting pharmaceuticals as well as the principles of the Internet of Things and the arising Physical Internet.

The most important improvement to be brought about by iTUs in contrast to existing solutions is the concept of smart shipping containers. The main advantages achieved with the indicated methods with respect to automatic routing, authentication and authorization of iTUs can be summarized as follows:

- Transport units can undeniably and justifiably be associated with accredited known consignors.
- The transport units' safety and security are guaranteed along entire supply chains.
- Less or even no personnel, who might manipulate the units, may be permitted in their surroundings.
- Confidentiality and security of the data flow accompanying the transport units is guaranteed.

References

1. IATA Homepage: IATA Cargo Strategy. In: IATA Cargo Strategy (2015). https://www.iata.org/whatwedo/cargo/Documents/cargo-strategy.pdf
2. IATA Homepage: Temperature Control Regulations. In: IATA Perishable Cargo Regulations (PCR) manual (2017). http://www.iata.org/publications/store/Pages/temperature-control-regulations.aspx
3. Embedded Homepage: Embedded – cracking the code to system development (2017). http://www.embedded.com/development/mcus-processors-and-socs
4. European Commission: Commission Regulation (EU) No 185/2010 of 4 March 2010 laying down detailed measures for the implementation of the common basic standards on aviation security (2010). https://eur-lex.europa.eu/LexUriServ/LexUriServ.do?uri=CONSLEG:2010R0185:20130711:EN:PDF
5. Montreuil, B.: Toward a physical internet: meeting the global logistics sustainability grand challenge. Logist. Res. **3**, 71–87 (2011). https://doi.org/10.1007/s12159-011-0045-x
6. Montreuil, B., Meller, R.D., Ballot, E.: Physical internet foundations. In: Borangiu, T., Thomas, A., Trentesaux, D. (eds.) Service Orientation in Holonic and Multi Agent Manufacturing and Robotics. Studies in Computational Intelligence. SCI, vol. 472, pp. 151–166. Springer, Heidelberg (2013). https://doi.org/10.1007/978-3-642-35852-4_10

7. ITU Homepage: Series Y: Global Information, Infrastructure, Internet Protocol Aspects and Next-Generation Networks, Internet of Things and Smart Cities, Next Generation Networks – Frameworks and functional architecture models, Semantics based requirements and framework of the Internet of things, Recommendation ITU-T Y.4111/Y.2076 (2016). https://www.itu.int/rec/T-REC-Y.2076-201602-I/en
8. Atzori, L., Iera, A., Morabito, G.: The internet of things: a survey. Comput. Netw. **54**(15), 2787–2805 (2010). https://doi.org/10.1016/j.comnet.2010.05.010
9. Xiao, L., Wang, Z.: Internet of things: a new application for intelligent traffic monitoring system. JNW **6**, 887–894 (2011). https://doi.org/10.4304/jnw.6.6.887-894
10. Brun, Y., Serugendo, G.D., Gacek, C., Giese, H., Kienle, H.M., Litoiu, M., Müller, H.A., Pezzè, M., Shaw, M.: Engineering self-adaptive systems through feedback loops. In: Cheng, B.H.C., de Lemos, R., Giese, H., Inverardi, P., Magee, J. (eds.) Software Engineering for Self-Adaptive Systems. LNCS, vol. 5525, pp. 48–70. Springer, Heidelberg (2009). https://doi.org/10.1007/978-3-642-02161-9_3
11. Seiger, R., Huber, S., Heisig, P., Assmann, U.: Enabling self-adaptive workflows for cyber-physical systems. In: Schmidt, R., Guédria, W., Bider, I., Guerreiro, S. (eds.) Enterprise, Business Process and Information Systems Modeling. BPMDS 2016, EMMSAD 2016. LNBIP, vol. 248, pp. 48–70. Springer, Cham (2016). https://doi.org/10.1007/978-3-319-39429-9_1
12. Wombacher, A.: How physical objects and business workflows can be correlated. In: IEEE International Conference on Services Computing, pp. 226–233. IEEE Computer Society, Washington (2011). https://doi.org/10.1109/scc.2011.24
13. Backus, J.W.: The syntax and semantics of the proposed international algebraic language of the Zurich ACM-GAMM conference. In: Proceedings of the International Conference on Information Processing, pp. 125–132. UNESCO, Paris (1959)
14. Gašević, D., Djurić, D., Devedžić, V.: Model Driven Engineering and Ontology Development, 2nd edn. Springer, Heidelberg (2009)
15. Gumzej, R., Halang, W.A.: Avtomatizirana avtentikacija in avtorizacija transportnih enot znanih dostavljavcev, Slovenian patent SI 25020 A (2015)
16. Gumzej, R., Halang, W.A.: Automatisierte authentifizierung und autorisierung von transporteinheiten bekannter versender, German patent registration 10 2017 000 706.3 (2017)
17. Zhou, L., Lou, C.X., Chen, Y., Xia, Y., Li, P.: Multi-agent-based smart cargo tracking system. IJHPCN **8**(1), 47–60 (2015)
18. European Commission: Regulation (EC) No 648/2005 of the European Parliament and the Council of 13 April 2005 amending Council Regulation (EEC) No 2913/92 establishing the Community Customs Code (2005). https://eur-lex.europa.eu/LexUriServ/LexUriServ.do?uri=OJ:L:2005:117:0013:0019:EN:PDF
19. Abdelgawad, A., Mahmud, A., Yelamarthi, K.: Butterworth filter application for structural health monitoring. IJHCR **7**(4), 15–29 (2016). https://doi.org/10.4018/IJHCR.2016100102

Feature Selection Method Based on Correlation Tree

Prajak Yapila[1(✉)] and Thanunchai Threepak[2]

[1] Defence Engineering, Faculty of Engineering,
King Mongkut's Institute of Technology Ladkrabang, Bangkok, Thailand
61601095@kmitl.ac.th

[2] Department of Computer Engineering, Faculty of Engineering,
King Mongkut's Institute of Technology Ladkrabang, Bangkok, Thailand
thanunchai.th@kmitl.ac.th

Abstract. Machine learning is one of techniques adapted to detect intrusion for cyber security. One of importance techniques to find anomaly is classification. But classification with huge dataset has the resources and time consumption. Feature selection is choice to reduce the data dimension to improve processing performance. In this paper, we introduce the new feature selection method that selects some fields of data set using position of each feature in correlation tree. Then, the result from the correlation tree feature selection of KDDCUP'99 data set are compared with two feature selection technique, correlation of coefficient (CC-type) and BFS by using three reference classifier, Decision Tree (DT), Random Forest (RF), and Naive Bayes (NB).

Keywords: KDDCup'99 · Machine learning · Feature selection · IDS · Cyber security

1 Introduction

At present, information security is most concern in every information system. Intrusion detection system comes to help administrator to detect intrusions or attacks. Classification is the one of machine learning technique behind intrusion detection to automate learning and finding intrusion. But, classification with big data are overwhelm system resource and performance while dimension reduction like PCA and feature selection technique come to reduce this problem by reduce data before classification process. There are many features selection techniques such as correlation based feature selection [1, 2], entropy based such as mutual information [3] and information gain [4–6], BFS algorithm [7, 8], Chi-square [4, 9, 10], Euclidean distance of the feature and class [11], and Wrapper Method feature and Filter Method [12] that work well to reduce data set.

This paper introduces a new feature selection method based on correlation tree. The concept behind this work comes from the assumption that, we can reduce two highly self-correlated features by choosing the higher one that correlates to class feature. The proposed algorithm and operation principle, such as KDDCUP'99 data set, correlation and covariance calculation are shown in Sect. 2. Section 3 show the experimental result and discuss performance measurement result. Finally, we conclude in Sect. 4.

P. Meesad and S. Sodsee (Eds.): IC^{2}IT 2020, AISC 1149, pp. 70–78, 2020.
https://doi.org/10.1007/978-3-030-44044-2_8

2 Operation Principle

2.1 KDDCUP'99 Data Set

KDDCUP'99 is the data set used for evaluation of anomaly detection method. This data is built based on DARPA'98 IDS evaluation program. The KDD training dataset consists of approximately 4,900,000 single connection vectors, each of which contains 42 features and is labeled as either normal or an attack, with exactly one specific attack type. The training set contains a total of 22 training attack types. Additionally, testing set includes an additional 17 attack types. Therefore, there are 39 attack types that are included in the testing set and these attacks can be classified into one of the four main classes;

- DOS: Denial of Service attacks.
- Probe: another attack type sometimes called Probing.
- U2R: User to Root attacks.
- R2L: Remote to Local attacks.

Features of KDDCUP'99 data set are shown in Table 1. Feature 0 to feature 40 is information about connection and feature 41 is the result from the detection process of intrusion detection system.

Table 1. KDDCUP'99 dataset.

No	Feature name	Description
0	Duration	Length (number of seconds) of connection
1	Protocol type	Type of protocol, e.g. tcp, udp, etc.
2	Service	Network service e.g., http, telnet, etc.
3	Flag	Normal or error status of connection
4	src_ bytes	Number of bytes from source to destination
5	dst_ bytes	Number of bytes from destination to source
6	Land	if connection is from/to the same host/port = 1; if not = 0
7	Wrong fragments	Number of wrong fragments
8	Urgent	Number of urgent packets
9	Hot	Number of hot indicators
10	num_failed_logins	Number of failed login attempts
11	Logged in	If successfully logged in = 1; if not = 0
12	num_compromised	Number of compromised conditions
13	Root shell	If root shell is obtained = 1; if not = 0
14	su attempted	If su root command attempted = 1; if not = 0
15	num_root	Number of root accesses
16	num_file_creations	Number of file creation operations
17	num_shells	Number of shell prompts

(*continued*)

Table 1. (*continued*)

No	Feature name	Description
18	num_access_files	Number of operations on access control files
19	num_outbound_cmds	Number of outbound commands in an ftp session
20	is_hot_login	If the login belongs to the hot list = 1; if not = 0
21	is_guest_login	If the login is a guest login = 1; if not = 0
22	Count	Number of connections to the same host as the current connection in the past two seconds
23	srv_count	Number of connections to the same service as the current connection in the past two seconds
24	serror rate	% of connections SYN errors
25	rerror rate	% of connections beside REJ errors
26	srv serror rate	% of connections beside SYN errors
27	srv rerror rate	% of connections beside REJ errors
28	same srv rate	% of connections to the same service
29	diff srv rate	% of connections to different services
30	srv diff host rate	% of connections to different hosts
31	dst host count	Number of connections having the same destination host
32	dst host srv count	Number of connections having the same destination host and using the same service
33	dst host same src rate	% of connections having the same destination that use the same service
34	dst host srv rate	% of connections having different hosts on the same system
35	dst host same srv port rate	% of connections having a system with the same source port
36	dst host srv diff host rate	% of connections having the same service coming from different hosts
37	dst host serror rate	% of connections having a host with an serror
38	dst host srv serror rate	% of connections having a host and specified service with an serror
39	dst host serror rate	% of connections having a host with an RST error
40	dst host srv serror rate	% of connections having a host and specified service with an RST error
41	Connection type	Types of attack

2.2 Pearson's Correlation and Covariance

Correlation is a statistical relationship between two variables. Two data sets that have high correlation value means these two data are highly related. A correlation can tell relation of two data sets. When items in a data set A and data set B move in the same direction, the correlation between A and B is positive. In other hand, when items in a data set A and data set B is moving in the opposite direction, the correlation between A and B is negative. When data are unrelated, data set A increases, but data in data set B decrease, correlation can also be zero. To calculate the correlation value, the correlation value of variable X and variable Y are displayed in the Eq. (1)

$$Corr(X, Y) = \frac{Cov(X, Y)}{Stdev(X) \times Stdev(Y)} \quad (1)$$

Covariance of X and Y are in covariance formula. Covariance is a statistical value that explains the relationship between the movements of two variables. When two data set tend to move together, they are positive covariance; when they move inversely, the covariance is negative. The covariance of the data set X and dataset Y are shown in the Eq. (2)

$$Cov(X, Y) = \frac{\sum (x - \bar{x}) * (y - \bar{y})}{N} \quad (2)$$

2.3 Correlation Tree Feature Selection

Feature selection is processed to reduce data set size. This procedure helps the classifier to process fewer data. This means using fewer resources and time. The good feature selection function produces a data set that remains a good score in performance measures such as accuracy, detection rate and false alarm rate Our algorithm is one of feature selection technique that use both correlations with classification and correlation with other feature to produce the tree of correlation. Then, select the feature from that tree to produce result data set for the classifier. The algorithm to produce correlation tree (CT) and correlation-tree list (CTL) descried in Table 2.

Table 2. Algorithm to generate correlation tree (CT) and correlation-tree list (CTL).

Step	Process/Sub-process
1	**Prepare sorted-positive correlation features list**
	• Separate data set into features list (FL) and class (c)
	• For each x in FL: if Corr(x, c) < 0 then remove x from FL
	• Sort FL in descending order by using Corr(t, c) value where t is each item in FL
2	**Construct correlation tree**
	• Initial correlation tree (CT) by one default root node (Node_R)
	• For each x in FL: - find y in FL where Corr(x, y) return maximum correlation - if y does not in CT then add x as child node of Node_R else add x as child node of y
3	**Generate correlation tree list (CTL)**
	• Initial the empty correlation-tree list (CTL)
	• For each tree layer(L) in CT: P = list of features in L Sort P in descending order by using Corr(x, c) value where x is each item in P Push each items from P into CTL

3 Experiments and Analysis

3.1 Generate Correlation Tree and Feature List for KDDCUP'99 Data Set

KDDUP'99 data set contains 42 features that identify characteristic of network connection, normal connection or some kind of attack. To select significant features using the proposed method, we first use the algorithm in Table 2 to find a list of features that have positive correlation with class. Then construct correlation tree (CT) of remaining features in the features list. Finally, we generate CTL of suitable features.

In the first step, feature 41 in KDDCUP'99 dataset is identified as a class feature and other features are identified as items in features list. After calculating correlation with class (feature 41), remove negative correlation, and sort by descending order, feature list finally contain 10 features, 23, 22, 1, 35, 32, 33, 28, 31, 2, and 4.

Then, each feature in feature list uses to construct correlation tree. The process starts with creating Node_R as default root node of correlation tree. Each item in CTL process sequentially. For example, Maximum correlation feature of feature 23 is feature 22 but the correlation tree doesn't have feature 22, then feature 23 is the child node of Node_R. Next, maximum correlation of feature 22 is feature 23 and feature 23 is in the tree, then feature 22 is the child node of feature 23. The sequence to generate correlation tree process until the end of the feature list. Then, we get the correlation tree in Fig. 1.

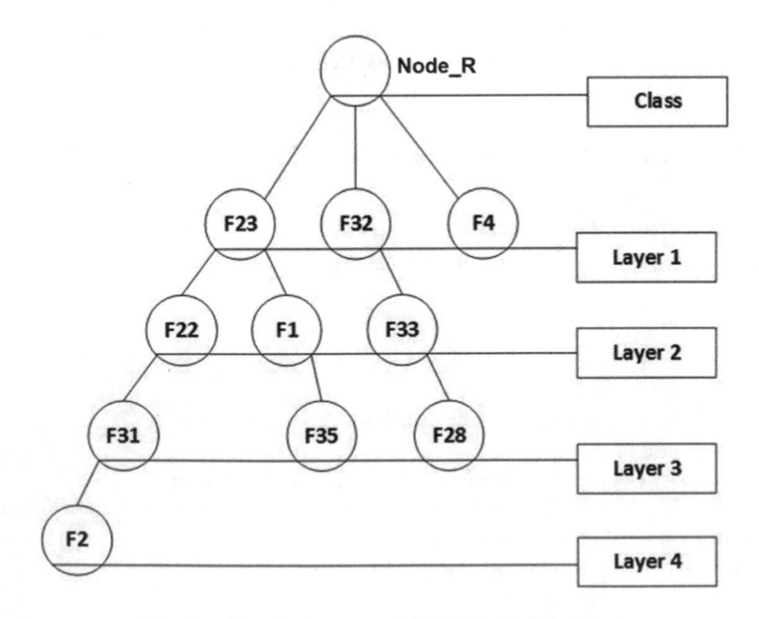

Fig. 1. Correlation tree of KDDCUP'99 features.

In the last phase, in each level of correlation tree, features are sorted by value of correlation with class. And we get the list of features rank by correlation tree technique are 23, 32, 4, 22, 1, 33, 31, 35, 28, and 2.

3.2 Data Set from CC and BFS Feature Selection

This experiment compares the correlation tree method with other two feature selection techniques, CC and BFS. Correlation Coefficient (CC) method generates a feature sequence by calculates the correlation value of each feature with the class and then sort in descending order. In the other hand, BFS use the best first search algorithm to generate feature list. The result set of KDDCUP'99 features from CC, BFS and Correlation Tree method shown in Table 3.

Table 3. Feature sequence of CC, BFS and Correlation tree.

Feature selection	Example
CC	23, 22, 1, 35, 32, 33, 28, 31, 2, 4
BFS	2, 4, 5, 7, 22, 23, 29, 34, 35, 36
Correlation tree	23, 32, 4, 22, 1, 33, 31, 35, 28, 2

3.3 Result and Discussion

To identify operation performance, Accuracy, the ratio the number of samples correctly classified in test data and number of samples in test data, is used to measure performance of each classification and evaluate the feature selection method. In this experiment, we use KDDCUP'99 data set with 10 to 2 features from three feature selection process, CC, BFS and CT method and compare accuracy result of 3 clustering algorithms. Clustering algorithms use in this test is both trees-based such as decision tree classification algorithm and random forest, and probability-based like naïve-Bayes classification.

Experiment 1: Accuracy of Decision Tree Classification Using CC, BFS and CT Feature Selection

Figure 2 compares the accuracy of decision tree classification of the 3 data set from CC, BFS and CT. It's clear that data set from CT and BFS provide higher accuracy than the data set from CC when classified using greater than three features. Through CT and BFS, CT has a little higher accuracy than BFS when using 4-8 features.

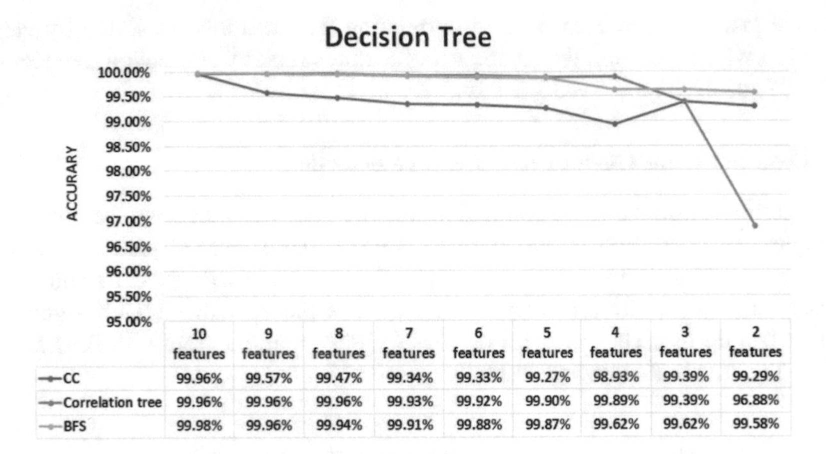

	10 features	9 features	8 features	7 features	6 features	5 features	4 features	3 features	2 features
CC	99.96%	99.57%	99.47%	99.34%	99.33%	99.27%	98.93%	99.39%	99.29%
Correlation tree	99.96%	99.96%	99.96%	99.93%	99.92%	99.90%	99.89%	99.39%	96.88%
BFS	99.98%	99.96%	99.94%	99.91%	99.88%	99.87%	99.62%	99.62%	99.58%

Fig. 2. Accuracy compare of CC, Correlation tree, and BFS feature selection in decision tree classification method.

Experiment 2: Accuracy of Naïve Bayes Classification Using CC, CT and BFS Feature Selection
Naïve-Bayes classification accuracy of CC, BFS and CT data set shown in Fig. 3. This show CT has higher accuracy than BFS and CC in all features set.

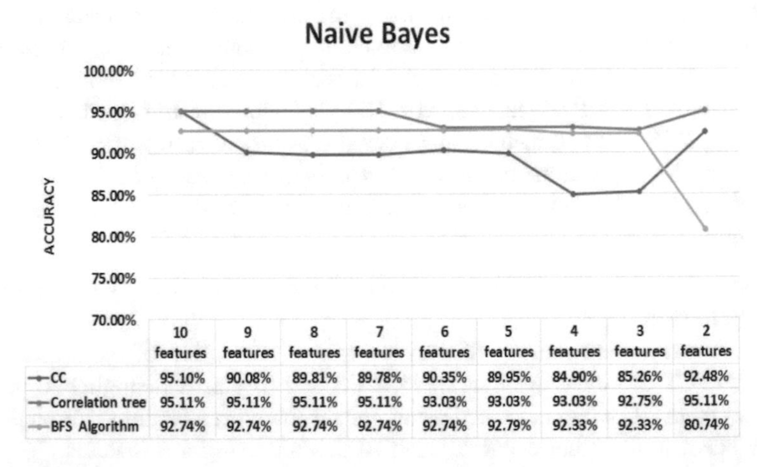

	10 features	9 features	8 features	7 features	6 features	5 features	4 features	3 features	2 features
CC	95.10%	90.08%	89.81%	89.78%	90.35%	89.95%	84.90%	85.26%	92.48%
Correlation tree	95.11%	95.11%	95.11%	95.11%	93.03%	93.03%	93.03%	92.75%	95.11%
BFS Algorithm	92.74%	92.74%	92.74%	92.74%	92.74%	92.79%	92.33%	92.33%	80.74%

Fig. 3. Accuracy compare of CC, Correlation tree, and BFS feature selection in Naïve Bayes classification method.

Experiment 3: Accuracy of Random Forest Classification Using CC, CT and BFS Feature Selection
Figure 4 compares the accuracy of random forest classification of the 3 data set from CC, BFS and CT. BFS has better accuracy when using 10 to 9 features and 3-2 features. CT has good performance when using 8 to 4 features.

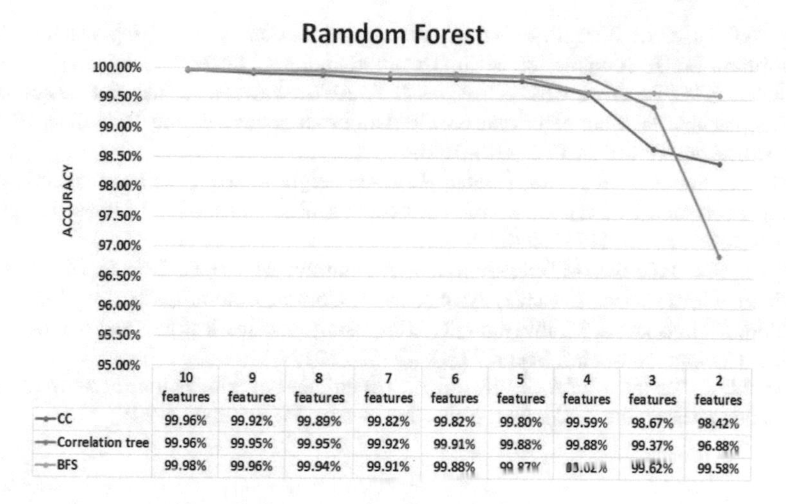

Fig. 4. Accuracy compare of CC, Correlation tree, and BFS feature selection in random forest classification method.

4 Conclusion

In this paper, we propose a method to select the feature in dataset by using correlation tree. The experimental results show that feature set selected from correlation tree has better results than CC and BFS in probability-based classification like Naïve Bayes. But in tree-based classification like CC and BFS, CT (Correlation Tree) feature selection has better result when using several numbers of features [13].

References

1. Mohammadi, S., Mirvaziri, H., Ghazizadeh-Ahsaee, M., Karimipour, H.: Cyber intrusion detection by combined feature selection algorithm. J. Inf. Secur. Appl. **44**, 80–88 (2019)
2. Eid, H.F., Hassanien, A.E., Kim, T.-H., Banerjee, S.: Linear correlation-based feature selection for network intrusion detection model. In: The International Conference on Advances in Security of Information and Communication, pp. 240–248 (2013)
3. Rachburee, N., Punlumjeak, W.: Big data analytics: feature selection and machine learning for intrusion detection on microsoft azure platform. J. Telecommun. Electron. Comput. Eng. **9**(1–4), 107–111 (2017)
4. Selvamani, D., Selvi, V.: A comparative study on the feature selection techniques for intrusion detection system. Asian J. Comput. Sci. Technol. **8**(1), 42–47 (2019)
5. Naganhalli, N.S., Terdal, S.: Network intrusion detection using supervised machine learning technique. Int. J. Sci. Technol. Res. **8**(09), 345–350 (2019)
6. El-Sappagh, S., Mohammed, E.-S., AlSheshtawy, T.A.: Multi-layer classifier for minimizing false intrusion. Int. J. Netw. Secur. Appl. **11**(03), 43–52 (2019)
7. Gunduz, S.Y., Çeter, M.N.: Feature selection and comparison of classification algorithms for intrusion detection. Anadolu Univ. J. Sci. Technol. A – Appl. Sci. Eng. **19**(1), 206–218 (2018)

8. Ugochukwu, C.J., Bennett, E.O.: An intrusion detection system using machine learning algorithm. Int. J. Comput. Sci. Math. Theory **4**(1), 39–47 (2018)
9. Othman, S.M., Ba-Alwi, F.M., Alsohybe, N.T., Al-Hashida, A.Y.: Intrusion detection model using machine learning algorithm on big data environment. J. Big Data **5**(1), 34 (2018). https://doi.org/10.1186/s40537-018-0145-4
10. Devi, K., Sukumar, R., Suresh Babu, R.: Lightweight network intrusion detection system using chi-square and cuckoo search optimization algorithms with decision tree classifier. Carib. J. Sci. **53**(1), 117–129 (2019)
11. Priyadarsini, P.I., Sai, M.S.S., Suneetha, A., Santhi, M.V.B.T.: Robust feature selection technique for intrusion detection system. Int. J. Control Autom. **11**(2), 33–44 (2018)
12. Malhotra, H., Sharma, P.: Intrusion detection using machine learning and feature selection. Int. J. Comput. Netw. Inf. Secur. **11**(4), 43–52 (2019)
13. Sakr, M.M., Tawfeeq, M.A., El-Sisi, A.B.: An efficiency optimization for network intrusion detection system. Int. J. Comput. Netw. Inf. Secur. **11**(10), 1–11 (2019)

Thai Words Segmentation Using an Unsupervised Learning Technique

Jirapon Sunkpho[1(✉)] and Markus Hofmann[2]

[1] Thammasat University, Bangkok 10400, Thailand
jirapon@tu.ac.th
[2] Technological University Dublin, Blanchardstown, Ireland

Abstract. Word Segmentation or Tokenization is the process of determining the best likely sequence of words from a sequence of text. For Thai language, word segmentation is not a trivial task as words and sentences in Thai are written continuously without any spaces or delimiters. Most techniques for word segmentation, especially when using machine learning, requires manually tagged data where words begin and end as a training dataset. In this study, an unsupervised machine learning technique that does not require the use of manually tagged data was developed. The technique involves breaking input text into syllables and then uses Genetic Algorithms (GA) to merge the syllables back into words. GA identifies the best segmentation of words by minimizing word distance which is the novel concept developed in this study. It is the sum of all syllable distances of every pair of syllables within a word. The syllable distance is the measure of how far apart each pair of syllables is in a document. The implementation was done using Python and achieves 70% accuracy (F1 measure) while using a 100k untagged words training dataset. The performance also improves with more training data and some tuning of GA parameters.

Keywords: Word segmentation · Unsupervised learning · Genetic Algorithms

1 Introduction

1.1 Background

Word segmentation is the process of determining the best likely sequence of words from an input text. It is one of the first tasks in performing text analytics such as document classification, sentiment analysis, or chatbots. For most Latin-based languages such as English or German, word segmentation is rather trivial because words can be separated by white spaces or other delimiters, e.g., semi-colon, comma, and period. However, for some languages like Chinese, Japanese, and Thai, words and sentences are written continuously without any spaces or delimiters making segmentation of words much more difficult. Several works on word segmentation of Thai language have been done since the 80s [2, 21]. However, due to the complexity of the problem, it is still an on-going research area [4, 10, 13, 16].

Most current techniques for Thai word segmentation are based on machine learning techniques. Supervised machine learning has been used to perform Thai word segmentation tasks [3, 9, 10, 21]. However, the supervised machine learning approach

P. Meesad and S. Sodsee (Eds.): IC^{2}IT 2020, AISC 1149, pp. 79–92, 2020.
https://doi.org/10.1007/978-3-030-44044-2_9

requires training corpus that has to be prepared by human experts. Preparing and maintaining this training corpus can be expensive. In order to avoid this problem, unsupervised machine learning techniques, which requires no tagging or labelling on where words begin and end, is becoming more attractive. Moreover, recent studies have shown that unsupervised techniques have been successfully applied to languages like Chinese and Japanese [1, 2, 11, 24, 26].

1.2 Literature Review

Thai is the official language of Thailand and it is the language for 65 million Thai citizens. Thai language is mainly used in the Indochina sub-region, e.g., Cambodia, Myanmar, Laos, Thai, and Vietnam [14]. It is written from left to right in non-segmented form, i.e., there are no spaces between words. There are 44 consonants, 21 vowels, and 4 tone markers. Vowels can be written in front of, at the bottom of, at the end of, or on the top of consonants [3]. Moreover, there is no period at the end of the sentence. Finally, there is no uppercase or lowercase text as every character has only one form.

To segment Thai words programmatically, one cannot rely on using spaces or commas to separate words. Rather, an algorithm designed to perform word segmentation task is needed. There are three main approaches in performing word segmentation including rule-based, dictionary-based, and statistical-based approaches.

Early efforts for Thai word segmentation rely on a rule-based approach in which rules that describes how words are formed are generated [6]. The algorithm then tries to match the input text with one of these rules. If matched, it returns the input text as a word. However, the problem of the rule-based approach is that it is impossible to create all the rules that represent all the Thai words since there can be thousands of such rules. Instead of creating all rules for Thai words, it was used to represent syllables which are small unit of a word. Since Thai words are the combination of one or more syllables, it is more practical to write rules to represent Thai syllables [12].

Dictionary-based approach: instead of creating rules to match all possible words, input text is compared with words stored in the dictionary (database) in order to determine if there are any matches between the input text and stored words. Two problems arise with this approach: 'unknown word' and 'ambiguity problem' [9]. 'Unknown word' occurs when word(s) in the input text do not exist in the dictionary, therefore, the algorithm could not find the matching word [9]. Theeramunkong and Usanavasin [23] found in their study that the accuracy drops to 48% for dictionary-based approach when the data contains 50% of unknown word. 'Ambiguity problem', on the other hand, occurs when there is more than one way to segment words [9]. One classic example is 'ตากลม' which can be segmented as either ['ตา (eye)' 'กลม (round)'] or ['ตาก (dry)' and 'ลม (wind)'] depending on the context of the sentence.

Dictionary-based approach, instead of creating rules to match all possible words, input text is compared with words stored in the dictionary (database) in order to determine if there are any matches between the input text and stored words. Two problems arise with this approach: 'unknown word' and 'ambiguity problem' [9]. 'Unknown word' occurs when word(s) in the input text do not exist in the dictionary, therefore, the algorithm could not find the matching word [9]. Theeramunkong and Usanavasin [23] found in their study that the accuracy drops to 48% for dictionary-based approach when the data contains 50% of unknown word. 'Ambiguity problem', on the other hands, occurs when there are more than one way to segment words [9]. One classic example is 'ตากลม' which can be segmented as either ['ตา(eye)' 'กลม(round)'] or ['ตาก(dry)' and 'ลม (wind)'] depending on the context of the sentence.

As rule-based and dictionary-based methods have their own problems, many researchers have applied statistical-based or machine learning techniques to word segmentation problem [3, 9, 10, 16, 23]. Most machine learning techniques used for the Thai word segmentation problem are supervised learning techniques in which training data has to be manually prepared for the machine learning algorithm to learn from. Although, supervised machine learning techniques have better performance than other approaches, they require a tagged corpus that was manually prepared by the language experts. However, in some circumstances, tagged corpus might not be readily available, or it is expensive to acquire [14].

Several researchers were trying to perform word segmentation tasks with unsupervised machine learning techniques without the use of tagged training datasets [1, 2, 5, 7, 11, 18, 20, 26]. Chen et al. [7] suggests there are two broad categories of unsupervised word segmentation: boundary prediction and word recognition. For boundary prediction, the algorithm predicts whether the boundary between two language units is a word boundary. Word recognition methods, in contrast, focus on recognizing word units using induction-based or language model-based.

Early efforts for Thai word segmentation without the use of tagged training datasets was discussed in [2]. In this approach, word segmentation was treated as a two-processes problem: syllable segmentation and syllable merging. During the syllable segmentation process, text was segmented into syllables using rule-based techniques. Adjacent syllables were then merged back into a word by determining whether a particular syllable boundary is the end of word using so-called 'collocation strength.' The collocation strength is calculated from the ratio of the probability that two syllables, x and y, occur together and the probability that x and y are separated by other syllables [2].

In our study, we first breaks text input into syllables using the rule-based approach. Once the text was broken into syllables, GA was used for merging the syllables back into words. In this process, GA tries to find the best segmentation of words from the input text so that the word distance is minimum. Word distance is the new concept developed in this study and it will be described in the next section.

2 Word Distance Concept

The smallest unit of a Thai word is defined as a syllable. Syllables can take many forms by combining consonants (c), vowels, and tonal (t) marks together. Figure 1 illustrates examples of different syllables combination rules. Each rule specifies where consonants and vowels can be placed. For example, the first rule can create words like 'จำ (remember)' or 'ป้า (I)'.

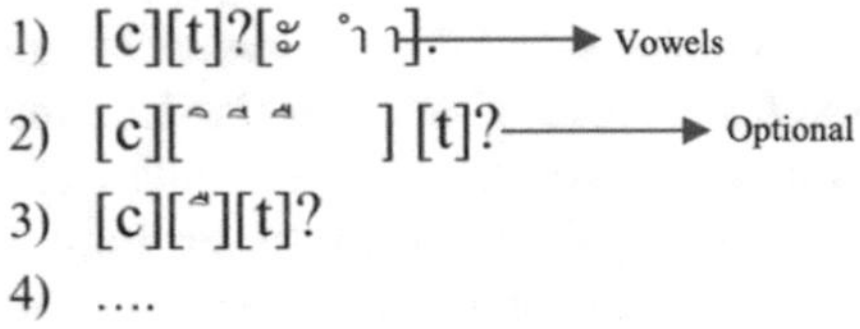

1) [c][t]?[ะ ำ า] → Vowels
2) [c][ิ ี ื] [t]? → Optional
3) [c][ื][t]?
4) ....

Fig. 1. Example of syllable formation.

In Thai, words can range from one to many syllables long. According to Wikipedia, a typical Thai word can be one syllable up until 18 syllables long [25]. Figure 2 illustrates the amount of Thai words according to the number of syllables. As one can see from Fig. 2, most Thai words have 2 syllables and words with 1–5 syllables account for 99% of all Thai words.

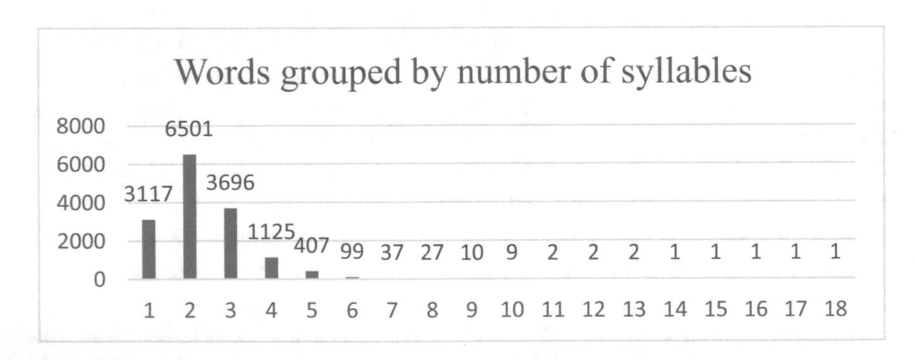

Fig. 2. Thai words according to the number of syllables.

In this study, we first introduced the concept of syllable distance. The syllable distance is defined as number of syllables +1, that appears between a pair of syllables of interest. For example, consider a sentence 'ฉัน (I)|เป็น(am)|นัก(a)|เรียน(student)', which has four syllables. The syllable distance between ฉัน(I) and เป็น(am), is then 1 since there is no syllable between ฉัน (I) and เป็น (am). The syllable distance for other pairs of syllables in this sentence can be calculated as illustrated in Fig. 3. The lower the syllable distance is, the more likely it is these syllables appear together, and thus are more likely to form a word.

I am a student

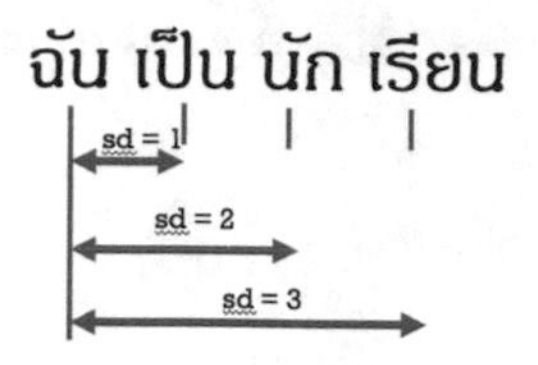

Fig. 3. Illustration of the syllable distance.

Once distance between all syllable pairs within this example sentence were calculated, a matrix of syllable distance and frequency matrix can be formed as illustrated in Fig. 4. Each element in the matrix represent the sum of distance between and the frequency of occurrence of a particular pair of syllables for a sentence 'ฉันเป็นนักเรียน (I am a student).'

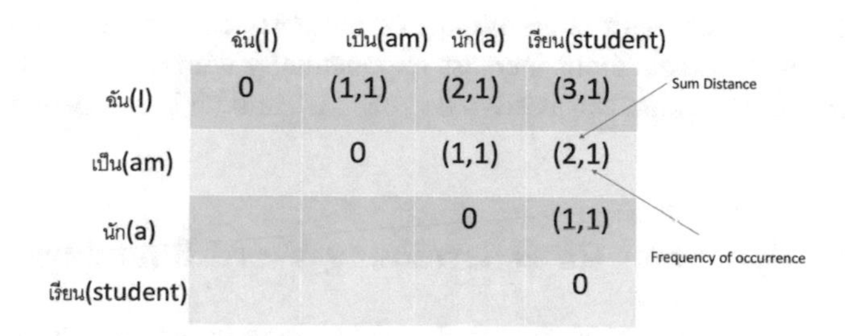

	ฉัน(I)	เป็น(am)	นัก(a)	เรียน(student)
ฉัน(I)	0	(1,1)	(2,1)	(3,1)
เป็น(am)		0	(1,1)	(2,1)
นัก(a)			0	(1,1)
เรียน(student)				0

Fig. 4. A syllable distance frequency matrix of a sentence.

The distance frequency matrix of Thai syllables can be created from a training dataset. Text input was first separated into syllables using rule-based technique described previously. The syllable distance was then calculated for each syllable pair encountered and thus the distance frequency matrix was formed. A typical distance frequency matrix is similar to the one illustrated in Fig. 5. Note that this kind of matrix is not symmetrical since the distance between two syllable may be different depending on the order of two syllables, i.e., the distance between 'ฉัน(I)' and 'เป็น(am)' is different from the distance between 'เป็น(am)' and 'ฉัน(I).'

		I ฉัน	am เป็น	a นัก	student เรียน
I	ฉัน	(5,2)	(4,3)	(8,3)	(6,2)
am	เป็น	(7,5)	(10,3)	(3,2)	(4,2)
a	นัก	(6,3)	(16,5)	(15,4)	(5,5)
student	เรียน	(14,4)	(20,6)	(25,7)	(10,5)

Sum of Distance Between syllables

Frequency of occurrence

Fig. 5. A typical distance frequency matrix generated from training data.

In order to determine whether a particular syllable sequence is a word, a word distance concept is introduced. Word distance is defined as the sum of average syllable distance for every pair of syllables within a word of interest. Word distance can be defined as

$$word\ \ distance = \sum\nolimits_{i=1}^{n} \left(sd_{i,i+1}\right) \tag{1}$$

Where

$sd_{i,i+1}$ is the syllable distance between s_i and s_{i+1}
n is the number of syllables

Once a word distance is calculated, the task of word segmentation is simply to identify the word boundaries within each sentence so that the sum of word distance of all possible words is minimum. One way to do that is to list all combinations of word boundaries and calculate the sum of word distance for every combination as will be explained in the next section. Calculating the sum of word distances for every combination of an input text with large number of syllables can be very time-consuming since the number of combinations increase exponentially with the number of syllables. Thus, the Genetic Algorithm was used to do such task and will be explained in the next section.

3 Genetic Algorithms for Determining Word Boundary

Using Genetic Algorithms (GA) for word segmentation is not new. There are works related to the use of GA for word segmentation [8, 11, 15, 17, 18]. Those works have demonstrated the applicability of GA to word segmentation problems with adequate accuracy and without too much complexity of the approach.

According to Kazakov and Manandhar [11], GA for word segmentation problems involves maintaining a set of individuals or population and applies the natural selection operators which include crossover and mutation repeatedly in order to generate new population. The fitness function is used to rank individuals to identify their survival in the next generation. Nguyen et al. [19] used Mutual Information as a fitness function for word segmentation problem of Vietnamese language. They achieved about 80% of acceptable level of segmentation judged by human experts.

An approach for Thai word segmentation using GA in this study is presented as follows:

Goal. Let S be the given sentence of n syllables: $S = s_1 s_2 \ldots s_n$. The GA goal is to find the most likely sequence of words in which the sum of word distance of every word, $w_k = s_i \ldots s_j$, in S is minimum.

Representation. The population is represented as a set of individuals where each individual representing one of the 2^{n-1} possible word combination where n is number of syllables in the input text. Each individual population is represented by a series of 0s and 1s in which each bit represents whether that particular syllable is a part of the word or it is the beginning of a new word. In other words, if the bit string is different from previous bit string, that particular syllable is the beginning of the word. For example, the 0 1 1 1 for a sentence ฉัน(I) เป็น(am) นัก(a) เรียน(student) is the representation of a 2-words sentence ฉัน(I) | เป็นนักเรียน(am a student) as illustrated below.

ฉัน(I) เป็น(am) นัก(a) เรียน(student)

0 1 1 1

ฉัน(I) | เป็น(am) นัก(a) เรียน(student)

Table 1 illustrates all possible combination of words within a sentence ฉัน(I) เป็น(am) นัก(a) เรียน(student). As one can see in Table 1, there are $2^{4-1} = 8$ possible word sequence that can be formed from a 4 syllables sentence. In order to determine the most likely sequence of words, the following steps in GA needs to be performed in order to determine the most likely combination

Table 1. All possible word combination of a sentence 'ฉัน(I) เป็น(am) นัก(a) เรียน(student)'.

ฉัน (I)	เป็น (AM)	นัก (A)	เรียน (Student)	Word combination
1	1	1	1	ฉันเป็นนักเรียน
0	1	1	1	ฉัน เป็นนักเรียน
1	0	1	1	ฉัน เป็น นักเรียน
1	1	0	1	ฉันเป็น นัก เรียน
1	1	1	0	ฉันเป็นนัก เรียน
1	0	0	1	ฉัน เป็นนัก เรียน
1	1	0	0	ฉันเป็น นักเรียน
1	0	1	0	ฉัน เป็น นัก เรียน

Initialization. During the initialization step, GA parameters were initiated such as population size, number of generations, cross-over fraction, and mutation rate. The initial populations were also created by randomly generated 0s and 1s bit strings with the length equals to the number of syllables in the sentence.

Cross-over. A standard one-point cross operation on bit strings was applied in this study. With cross-over operation, a pair of individuals in population pool were combined so that two new offsprings were created.

Mutation. An individual was also randomly selected according to mutation rate. A particular bit in an individual was altered from 1 to 0 or from 0 to 1. In this study, the initial mutation rate used is 0.05.

Selection. For each generation, top N individuals with the lowest fitness function (which is described in the following section) were selected for reproduction (crossover and mutation) of the next generation.

Fitness Function. To evaluate the individuals, the fitness score for each individual was calculated based on (1). Word distance is the measure of how likely a particular word of interest is actually a word. The concept of word distance was discussed in detail in previous section.

$$fitness(id) = fit(w_1, w_2, \ldots, w_n) = \sum wd(w_i) \tag{2}$$

where

wi is the ith word within a sentence
wd is word distance function
id is the population id.

Convergence. In order to determine when to stop the GA process, the best fitness score of the current population is checked whether it is converged. Otherwise, the GA process will continue until the maximum number of generations is reached.

4 Implementation and Experimentation

The implementation of word segmentation using the approach discussed in previous section is illustrated in Fig. 6. There are 3 main processes implemented with Python involved in the steps of syllable segmentation; distance frequency matrix calculation; and the GA to determine the best segmentation for the input text.

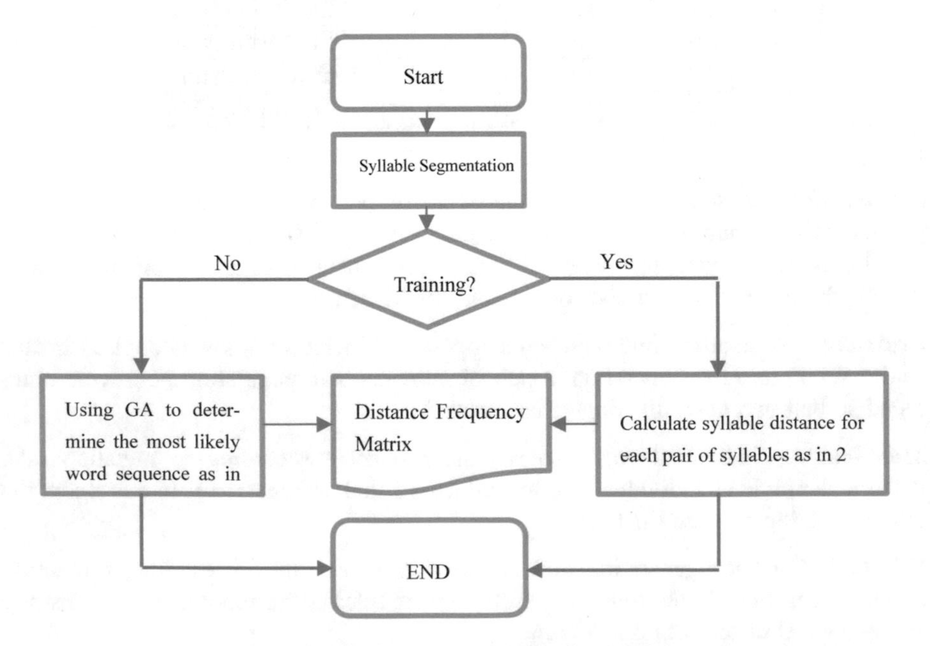

Fig. 6. Word segmentation implementation.

The first part of the implementation involves the syllable segmentation process. The syllable segmentation part was implemented using the Thai Language Toolkit (tltk) library in Python. This library has several functions involving manipulation of Thai text including syllable segmentation [22]. The module takes input text from training dataset 9 and turns it into sequence of syllables in order for the next module to create the distance frequency matrix.

The second part involves creating the distance frequency matrix from the sequence of syllables. This module calculates the syllable distance and the frequency of occurrence for each pair of syllables and generate the distance frequency matrix. This distance frequency matrix is required by the GA to compute word distances.

Finally, the last part of the implementation is the implementation of the GA and how it calculates the fitness function using the word distance concept. GA was used to identify the most likely sequence of word combinations that score the lowest word distance.

Since the method used in this study is an unsupervised technique, the dataset used to train the model, in fact, can be any text document. There are two sets of data used in the experiment. The first set of data is used during the model training. It was extracted from the public dataset called 'BEST' corpus [18]. The untagged data from BEST corpus contains 433,645 characters and 106,180 words and was used as a training dataset. Another set of data was used for testing the performance of the model. There are three types of documents including article, encyclopedia, and news that were extracted from [18]. Words within these documents were tagged by linguists. There are about 10,000 words in each document as illustrated in Table 2.

The training dataset was used to train the model to generate syllable distance frequency matrix. Then, three documents in the testing dataset were used e.g. encyclopedia, news, and articles for evaluation. Precision, recall, and F1-measure were used as evaluation criteria in this study. Precision, Recall, and F1 score are defined as follows:

$$Precision = \frac{Correct\ Words}{All\ segmentation\ produces} \tag{3}$$

$$Recall = \frac{Correct\ Words}{All\ Words} \tag{4}$$

$$F1 = \frac{2 * Precision * Recall}{(Precision + Recall)} \tag{5}$$

The initial result is illustrated in Table 2 with initial GA parameters of 30 initial populations, 100 iterations, and 0.05 mutation rate. The average F1 score is approximately 69.42%. The F1 performance for this model, when compared with other approaches, is still higher than the HMM approach in [3]. It also performs better than Nguyen's [19] work that employs GA for segmentation of Vietnamese text. The size of training data used for this model is 106,180 words compared to 113,404 words of manually tagged data were used in [3] and 534,330 in [9].

Table 2. Initial model performance.

Document type	Precision	Recall	F1	No. of words
Article	67.29	70.50	68.86	8,520
Encyclopedia	67.49	72.36	69.84	13,399
News	68.19	70.97	69.55	10,814

5 Causes of Mis-segmentation

Once the testing has been done, the result was then examined to see the causes of mis-segmentation or misclassification by the model. Even though, the overall performance is quite impressive given the amount of training data used compared with other studies in the literature. Some of the mis-segmentation needs to be examined so that the model could be improved.

There are two main causes for misclassification found in this study. The first one occurs when the model has not seen enough training data. For example, the model did not identify the word 'เกิดขึ้น (happen)' as one word but it identified 2 words: 'เกิด (birth)' and 'ขึ้น(up)'. This results as the word 'เกิดขึ้น(happen)' did not appear in the training dataset. As a result, the syllable pair 'เกิด (birth)' and 'ขึ้น(up)' does not exist in the distance frequency matrix, which in turn, makes the word distance higher than other word combinations. This type of misclassification can be alleviated by increasing the size of training data to increase the chance of seeing more words.

Another type of misclassification is more critical because the syllable pair may appear in the training dataset but the model still misclassifies the words. For example, a two-syllable word 'ร่วมกัน(together)' was misclassified as 2 words 'ร่วม(join)' and 'กัน(prevent).' This misclassification arose because the syllables 'ร่วม(join)' and 'กัน(prevent)' sometimes are next to each other and sometime are separated by other syllables, e.g., 'ร่วม(join)|ด้วย(too)|ช่วย(help)|กัน (prevent).' For this particular case, the syllables 'ร่วม(join)' and 'กัน(prevent)'were separated by other two syllables 'ด้วย (too)' and 'ช่วย (help)' making the syllable distance between 'ร่วม(join)' and 'กัน(prevent)' to increase and causing the word distance to increase. This type of problem is more difficult to solve but could also be alleviated with more training data as it increases the chance of encountering more correct segmentations.

6 Improving Accuracy of the Model

One of the reasons for mis-segmentation came from the fact that the text in the test set has not been seen by the model as discussed in previous sections. One way to improve the accuracy of the model is to train the model with a larger training data set in order for the model to see more syllable combinations. As a result, the model was trained again with a 200k words dataset extracted from [18]. The performance of the model was then illustrated in Table 3.

Table 3. Performance with more training data.

Document type	Precision	Recall	F1
Article	67.84	72.33	70.02 (68.86)
Encyclopedia	68.08	74.56	71.17 (69.84)
News	63.24	72.99	67.77 (69.55)

It is evident that with more training data, the performance of the model (F1 score) is better for article and encyclopedia document. However, the performance is slightly lower for the news document. Other ways to improve the model performance are to adjust the GA parameters in order to see the effects of these parameters on the model performance. Three parameters: number of iterations; initial population; and mutation rate, were examined to see the effects on the overall performance and keeping other parameters unchanged.

The first parameters for GA studied for the effect of performance is the number of iterations. For long sentence, e.g., 20 syllables sentence, there are $2^{20-1} = 524{,}288$ combinations of the word. In order for GA to examine these combinations thoroughly, it needs more iterations. From Table 4, the performance of the model seems to increase as the number of iteration increases.

Table 4. Performance with different GA iteration.

No of iteration	Precision	Recall	F1
50	66.33	70.00	68.11
100 (default)	67.29	70.50	68.86
200	67.84	71.45	69.60

The second parameter studied is the effect of the number of initial populations. Number of initial populations affects the performance of the model by increasing the chance of finding the optimal solution (segmented sentence with lowest word distance score) for the model. The results from Table 5 illustrate that the larger the initial population, the higher the performance for the model. However, increasing the initial population, the time required for the model to finish segmentation also increases.

Table 5. Performance with different initial population.

Initial population	Precision	Recall	F1
20	66.21	70.17	68.13
30 (default)	67.29	70.50	68.86
60	68.07	71.62	69.80

Finally, the last parameter for the GA model studied is the mutation rate. Higher mutation rate could increase the chance of finding optimal segmentation of words as it could help the model to escape local optimums. For this, the mutation rate varied from 5% to 20% as illustrated in Table 6. The optimal mutation rate found in this study is at 5% given other factors are constant. High mutation rate increases the probability of searching more combinations of segmentations. However, it prevents the population to converge to the optimum segment. In other words, too high mutation rate is similar to a random walk while too small mutation rate will only reach the local optimum.

Table 6. Performance with different mutation rates.

Mutation rate	Precision	Recall	F1
0.05 (default)	67.29	70.50	68.86
0.1	64.49	69.03	66.69
0.2	62.19	66.88	64.46

7 Summary and Conclusion

Most text analytics applications require the input text to be tokenized or segmented into words so that they can be processed by other text analytics methods. For Thai language, word segmentation is not trivial since Thai words and sentences are written continuously without any spaces or delimiters, thus, using space or delimiters to segment text into words is impossible. Most successful Thai text segmentation methods rely on the use of supervised machine learning techniques to train the model with a dataset that was prepared manually. However, preparing and maintaining training the dataset is an expensive task making unsupervised machine learning method more attractive.

In this study, the unsupervised machine learning approach for segmentation of Thai text without having to prepare tagged corpus manually was developed. The model first breaks text into syllables using a rule-based approach for segmentation. Once the text was broken into syllables, GA was used for merging the syllables back into words. In this process, GA tries to find the best segmentation of words from the input text so that the word distance is minimum. Word distance is the new concept developed in this study. It is the sum of all syllable distance of every pair of syllables within a word where the syllable distance is the measure of how far apart each pair of syllables is in a document. The concept of syllable distance is based on the assumption that syllables that are closer together are more likely to form a word.

The implementation was done using Python. The model illustrates promising results with approximately 70% accuracy while using only a 100k-word training dataset. The performance is even better than previous studies using supervised machine learning approach with larger training dataset. The model performance can be improved by training the model with larger training dataset as it increases chance of encountering more words. Several GA parameters were also examined, and it was found that with more GA iterations and initial population, the performance improved.

References

1. Ando, R.K., Lee, L.: Mostly-unsupervised statistical segmentation of Japanese Kanji sequences. Nat. Lang. Eng. **9**(2), 127–149 (2003)
2. Aroonmanakun, W.: Collocation and Thai word segmentation. In: Proceedings of the Fifth Symposium on Natural Language Processing & the Fifth Oriental COCOSDA Workshop, pp. 68–75 (2002)
3. Bheganan, P., Richi, N., Xu, Y.: Thai word segmentation with hidden Markov model and decision tree. In: Proceedings of the 13th Pacific-Asia Conference on Advances in Knowledge Discovery and Data Mining, Bangkok (2009)
4. Boonkwan, P., Supnithi, T.: Bidirectional deep learning of context representation for joint word segmentation and POS tagging. In: Le, N.T., van Do, T., Nguyen, N., Thi, H. (eds.) Advanced Computational Methods for Knowledge Engineering (2018)
5. Chang, J.S., Lin, T.: Unsupervised word segmentation without dictionary. In: The Association for Computational Linguistics and Chinese Language Processing (ACLCLP), pp. 355–359 (2003)
6. Chanyapornpong, S.: A Thai syllable separation algorithm. Master thesis, Asian Institute of Technology, Thailand (1983)
7. Chen, S., Xu, Y., Chang, H.: A simple and effective unsupervised word segmentation approach. In: Proceeding of the Twenty-Fifth AAAI Conference on Artificial Intelligence, San Francisco, USA (2011)
8. Detorakis, Z., Tambouratzis, G.: Applying a sectioned genetic algorithm to word segmentation. Pattern Anal. Appl. **13**(1), 93–104 (2010)
9. Haruechaiyasak, C., Kongyoung, S., Dailey, M.: A comparative study on thai word segmentation approaches. In: Proceedings of ECTI-CON (2008)
10. Jousimo, J.: Thai word segmentation with bi-directional RN (2017). https://sertiscorp.com/thai-word-segmentation-with-bi-directional_rnn
11. Kazakov, D., Manandhar, S.: Unsupervised learning of word segmentation rules with genetic algorithms and inductive logic programming. Mach. Learn. **43**, 121–162 (2001)
12. Khankasikarn, K., Muansuean, N.: Thai word segmentation a lexical semantic approach. In: Proceedings of the Tenth Machine Translation Summit (2005)
13. Kittinaradorn, R., Chaovavanich, K., Achakulvisut, T., Kaewkasi, C.: Deepcut (2018). https://github.com/rkcosmos/deepcut
14. Koanantakool, H.T., Karoonboonyanan, T., Wutiwiwatchai, C.: Computers and the Thai language. IEEE Ann. Hist. Comput. **31**(1), 46–61 (2009)
15. Lamprier, S., Amghar, T., Levrat, B., Saubion, F.: SegGen: a genetic algorithm for linear text segmentation. In: IJCAI 2007, Proceedings of the 20th International Joint Conference on Artificial Intelligence, Hyderabad, India, 6–12 January 2007 (2007)
16. Lapjaturapit, T., Viriyayudhakorn, K., Theeramunkong, T.: Multi-candidate word segmentation using bi-directional LSTM neural networks. In: Proceedings of the 11th International Conference on Embedded Systems and Intelligent Technology in cooperation with the 9th International Conference on Information and Communication Technology for Embedded Systems (ICESIT-ICICTES 2018), Khon Kaen, Thailand, pp. 30–35 (2018)
17. Mohammed, A., Karam, M., Hefny, H.: GA-based parameter optimization for word segmentation. Artif. Intell. Mach. Learn. J. **17**(1), 23–32 (2017)
18. Nectec. Annotated and Multimedia Corpus. National Electronics and Computer Technology Center. https://www.nectec.or.th/corpus/index.php?league=pm. Accessed 21 Nov 2019

19. Nguyen, T.V., Tran, H.K., Nguyen, T.T.T., Nguyen, H.: Word segmentation for Vietnamese text categorization: an online corpus approach. In: The 4th International Conference on Computer Sciences Research, Innovation and Vision for the Future (2006)
20. Peng, F., Schuurmans, D.: A hierarchical EM approach to word segmentation. In: Proceedings of the Sixth Natural Language Processing Pacific Rim Symposium (NLPRS 2001), Tokyo, Japan, November 2001 (2001)
21. Poowarawan, Y.: Dictionary-based Thai syllable separation. In: Proceedings of the Ninth Electronics Engineering Conference (1986)
22. PyPI: ttlk 1.2.1 Thai Language Toolkit. https://pypi.org/project/tltk/. Accessed 21 Nov 2019
23. Theeramunkong, T., Usanavasin, S.: Non-dictionary-based Thai word segmentation using decision trees. In: Proceedings of the First International Conference on Human Language Technology Research, San Diego, California, 18–21 March 2001, pp. 251–256 (2001)
24. Wang, H., Lepage, Y.: Unsupervised word segmentation using minimum description length for neural machine translation. In: The Association for Natural Language Processing (2018)
25. Wikipedia. Thai words by number of syllables (2019). https://en.wiktionary.org/wiki/Category:Thai_words_by_number_of_syllables
26. Zhikov, V., Takamura, H., Okumura, M.: An efficient algorithm for unsupervised word segmentation with branching entropy and MDL. Inf. Media Technol. **8**(2), 514–527 (2013)

A Framework for Designing and Evaluating Persuasive Technology in Education

Waransanang Boontarig(✉) and Charnsak Srisawatsakul

Faculty of Computer Science, Ubon Ratchathani Rajabhat University, Ubon Ratchathani, Thailand
Waransanang.b@ubru.ac.th, Charnsak@researcher.in.th

Abstract. Persuasive technology can influence people to change their behavior. It can be also useful in the education context. For example, it could be integrated with pedagogy for persuading students' attention in the classroom. Therefore, this research applies the principles of persuasive technology in a university classroom. There are three stages of the proposed framework used in this research. Firstly, the research collected users' requirements on the persuasive technology in education and list the top 10 most selected persuasive elements. Secondly, the research selected the applications that included elements from the first stage and apply in a classroom for one semester. Furthermore, this study collected data from the student for finding factors affecting the intention to adopt persuasive technology in the classroom. Lastly, the study evaluated users' performance by comparing the test scores from students who used persuasive technology with the students who did not use it. The results from multiple linear regression showed that primary task support and social support were significantly influencing students' intention to adopt persuasive technology in the classroom. The classrooms that used persuasive technology have better average scores than the class that did not. Possible future research is also discussed.

Keywords: Persuasive technology · Pedagogy · Users' requirements · Education technology

1 Introduction

Presently, most of the tertiary students are the generation Z (Gen-Zers; born between 1990–1999) [1]. They are also called digital natives because they were growing up with digital technology [2, 3]. Seemiller and Grace [4] explained that Gen-Zers described themselves as loyal, thoughtful, compassionate, open-minded, and responsible. The problems of this generation with education are that they may have less concentration in the classroom. Most of them learn best by doing and seeing than listening to the lecture [5]. Moreover, the average attention span of this generation is only around 8-s. [6] explained this effect as "acquired attention deficit disorder". One of the reasons that cause this problem is that they could easily use the smartphone to search for any knowledge from the internet. Therefore, [7] suggested that the methods of university education and technology have continued to emerge. Hence, the use of technologies in the education processes could increase the performance of the students in the classroom

P. Meesad and S. Sodsee (Eds.): IC^{2}IT 2020, AISC 1149, pp. 93–103, 2020.
https://doi.org/10.1007/978-3-030-44044-2_10

[8]. However, previous research did not pay much attention to the usage of persuasive technology in education [9]. Therefore, the purpose of this research is to develop a framework for using persuasive technology in education. Firstly, the research started with an extensive literature review on persuasive technology and its elements. Secondly, the study experiment to find the top ten elements which can persuade students in the classroom. Thirdly, the study collected data from the sample and finding the relationship between each element of persuasive technology and the intention to use persuasive technology in the classroom. Lastly, the proposed research framework was tested from two groups to compare the learning outcome between a group with and without persuasive technology.

The major results of this study will hopefully be a good first indication of interesting patterns and will, therefore, be used for our further higher number of participant's empirical research. It should also help the lecturer and other related professionals to use more effective persuasive technology, especially in the education context.

2 Literature Reviews

2.1 Persuasive Technology

Fogg [10] stated the persuasive technology can be defined as an interactive information technology designed for changing users' attitudes or behavior. At first, computers were not originally created to persuade. They were built for handling data such as processing, storing and retrieving. However, since computers have shifted from research labs onto the desktops and people are using it in daily life, they have become more persuasive by design. Today, it can act as persuaders and influencers that traditionally were filled by teachers, coaches, doctors, and sale people among others. Fogg [10] also proposed that there are three components of the computer that could be designed as persuasive technology, which are tools, media, and social actors. Later, He presented the eight steps in the process of design persuasive technology including 1. choose a simple behavior as a target 2. chooses a receptive audience 3. to find what is preventing the audience to perform the target behavior 4. choose an appropriate technology channel 5. find a relevant example of persuasive technology 6. imitate successful examples 7. test and iterate quickly 8. expand on the success [11]. Moreover, Oinas-Kukkonen and Harjumaa utilized the idea of persuasive technology by categorizing the principles of design strategies and guidelines to primary task support, dialogue support, system credibility support, and social support [12].

Persuasive technology concept has been applied in various domains such as healthcare, fitness, commerce, education, etc. For example, Edward et al. [13] studied persuasive in digital devices to motivate physical activity in adolescents. Kadomura et al. [14] developed a digital fork and a mobile interactive and persuasive game for a young child who is a picky eater to motivate their eating behavior. Alhammad and Gulliver [15] examined persuasive features used in e-commerce design. The results showed extensive use of persuasive features including dialogue support, credibility support, and primary task support.

2.2 Persuasive Technology in Education

Previous studies suggested that persuasive technology could be utilized to improve the learning outcomes of students. For example, Behringer et al. [16] applied the concept of persuasive technology to develop tools for the learner. They indicated that those tools were able to increase the positive attitude of students toward the education processes [16]. Likewise, Goh et al. [17] also used the principle of persuasive technology in the classroom by sending short message service (SMS) to students. The result showed that students who received a persuasive message performed better in the classroom than others. Filippou et al. [18] also investigated students' behaviors on academic performance. They suggested that persuasive technology could be applied to create a learning environment based on their study habits which could increase students' intention toward learning. Widyasari et al. [19] also integrated persuasive strategies and web 2.0 in online learning. The result showed that online learning by using a persuasive concept can improve the users' intention.

3 Proposed Research Framework

The proposed framework of this study shows in Fig. 1. Firstly, it shows how to select appropriate persuasive elements by gathering users' requirements from students. The persuasive technology elements based on Fogg [10] and categorized followed Oinas-Kukkonen et al. [12]. Secondly, design computing technology based on requirements gathered from the first step. The last step is to evaluate user performance of using persuasive technology which design based on users' requirements.

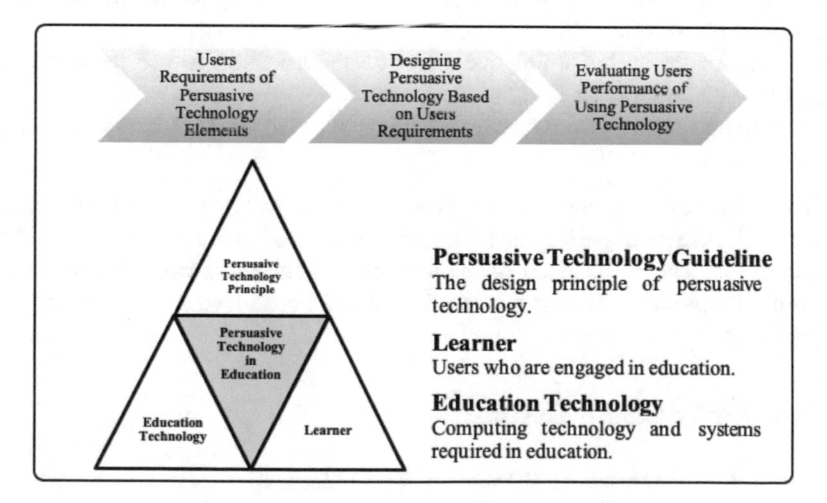

Fig. 1. The research framework for designing and evaluating persuasive technology in education.

4 Research Methodology

The preliminary study was gathering information from students in the faculty of computer science, Ubon Ratchathani Rajabhat University. There were three stages following the research framework. In the first stage, the questionnaire was used to find the top ten acceptance elements from persuasive technology. The design of the questionnaire was based on Fogg [10] and Oinas-Kukkonen et al. [12]. The questionnaire consists of two parts. The first part asked about demographic information. The second part focused on participants' requirements on persuasive technology in education. The questionnaire uses 5-point Likert scale (5 = strongly agree; 4 = agree; 3 = neutral; 2 = disagree; 1 = strongly disagree). The participants will be informed that 'starting the questionnaire indicates that this questionnaire is voluntary. A total of 87 questionnaires were collected. After a data screening process, 62 responses passed the screening and were used for analysis. The demographics information shows that there are 27 males and 35 females. The average age is 22 years old.

After data was collected from the first stage, this research presented an existing application relating to users' requirements. Then, at the second stage, the selected elements were categorized following Oinas-Kukkonen et al. [12] which were primary task support, dialogue support, system credibility support, and social support. Participants were asked to provide their answers by thinking that they were required to adopt an existing application as persuasive technology. This questionnaire also used a 5-point Likert scale. Therefore, the hypotheses of this study are:

H1: Primary Task Support has a positive effect on the intention to use persuasive technology in the classroom
H2: Dialogue Support has a positive effect on the intention to use persuasive technology in the classroom
H3: System Credibility has a positive effect on the intention to use persuasive technology in the classroom
H4: Social Support has a positive effect on the intention to use persuasive technology in the classroom

Finally, in order to confirm the proposed framework, this research evaluates learners' performance of participants. The performance of students was evaluated using an examination. The comparison of the test scores between current students and students from the previous semester without applying persuasive technology in class is also presented.

5 Data Analysis and Results

5.1 Users Requirements of Persuasive Technology Elements

The data from the first questionnaire were analyzed using mean and standard deviation in order to find the top ten most accepted elements of persuasive technology. The results of the top ten elements of persuasive technology were present in Table 1. Those ten elements considered as the most important elements that would improve the performance of the tertiary student in the classroom using persuasive technology.

Table 1. Top 10 elements of persuasive technology principles for education.

No.	Principles	Mean	Std. dev
1	Social learning	4.20	0.67
2	Social facilitation	4.15	0.81
3	Tailoring	4.13	0.85
4	Suggestion	4.10	0.86
5	Normative influence	4.07	0.80
6	Self-monitoring	4.03	0.86
7	Social role	4.00	0.75
8	Tunneling	3.98	0.80
9	Trustworthiness	3.98	0.89
10	Attractiveness, liking	3.98	0.86

5.2 Designing Persuasive Technology Based on Users Requirements

For the second stage, this study used Google classroom and Facebook as the applications for experiment with the students in the classroom because they have the elements that matched with the top ten elements of persuasive technology (Table 2).

Table 2. The selected persuasive elements with the description and example features.

No.	Principles	Description	Features
1	Social learning (social support)	A person will be more motivated to perform a target behavior if he or she can use computing technology to observe others performing the behavior and being rewarded for it [10, 12]	Google classroom and Facebook provide timeline features that users can generate content, observe and affiliate others
2	Social facilitation (social support)	People are more likely to perform a well-learned target behavior if they discern via the system that others are performing the behavior along with them [10, 12]	Google classroom and Facebook provide a timeline feature that users can see others performing
3	Tailoring (primary task support)	Information provided by computing technology will be more persuasive if it is tailored to the individual's needs, interests, personality, usage context, or other factors relevant to the individual [10, 12]	Google classroom and Facebook provide stream and post functions that users can reach the content they needed

(*continued*)

Table 2. (*continued*)

No.	Principles	Description	Features
4	Suggestion (dialogue support)	Computing technology will have greater persuasive power if it offers suggestions at opportunity moments [10, 12]	Google classroom and Facebook provide help center to support users when they need
5	Normative influence (social support)	Computing technology can leverage normative influence (peer pressure) to increase the likelihood that a person will adopt or will avoid performing a target behavior [10, 12]	Google classroom and Facebook provide a timeline feature that users can see others performing
6	Self-monitoring (primary task support)	Applying computing technology to eliminate the tedium of tracking performance or status helps people to achieve predetermined goals or outcomes [10, 12]	Google classroom provides a Classwork feature that users can track their works
7	Social role (dialogue support)	If a system adopts a social role, users will more likely use it for persuasive purposes [10, 12]	Google classroom and Facebook provide the comment and reply feature that users can ask for answers and exchange ideas
8	Tunneling (primary task support)	Using computing technology to guide users through a process or experience provides opportunities to persuade along the way [10, 12]	Google classroom and Facebook contain features for the teacher to deliver the content in sequential lessons and also provide a list of the activities
9	Trustworthiness (system credibility support)	Computing technology that is viewed as trustworthy (truthful, fair, and unbiased) will have increased powers of persuasion [10, 12]	Google classroom and Facebook are trustworthy as they well know the application
10	Attractiveness, liking (dialogue support)	A system that is visually attractive for its users is likely to be more persuasive [10, 12]	Google classroom and Facebook are a common use for the student which should attract them to use as learning tools

For this second stage, the research also verifies the users' perception of selected applications. The selected persuasive elements were group following the Oinas-Kukkonen et al.'s [12] concept, which divided into four categories: Primary Task Support, Dialogue Support, System Credibility Support, and Social Support. These four categories were used as factors affecting the intention to adopt persuasive technology using Multiple Linear Regression. The output presents the standardized regression coefficient, standard error of coefficient and the t-value for each estimate. As the t-value serves as a standard for defining the level of significance, it should greater than 1.96 at the 95% confidential level [20]. Figure 2 and Table 3 present the result of regression analysis. The behavioral intention to adopt persuasive technology in education accounts for 41.7% of the variance an intention. The analysis showed that there were statistically significant between Primary Task Support ($\beta = 0.648$, $p < 0.05$) and Social Support ($\beta = 0.406$, $p < 0.05$) the intention to adopt google classroom and Facebook as persuasive technology.

Table 3. Regression analysis result summary.

Dependent variable	R^2	Independent variables	β	Standard error of β	t	P
INT	0.417	**PTS**	**0.648***	**0.366**	**2.136**	**0.037**
		DS	−0.532	0.394	−1.539	0.129
		SCS	0.155	0.207	0.785	0.436
		SS	**0.406***	**0.243**	**2.433**	**0.018**

PTS (Primary Task Support); DS (Dialogue Support); SCS (System Credibility Support); SS (Social Support)

*Significant at 0.05 level

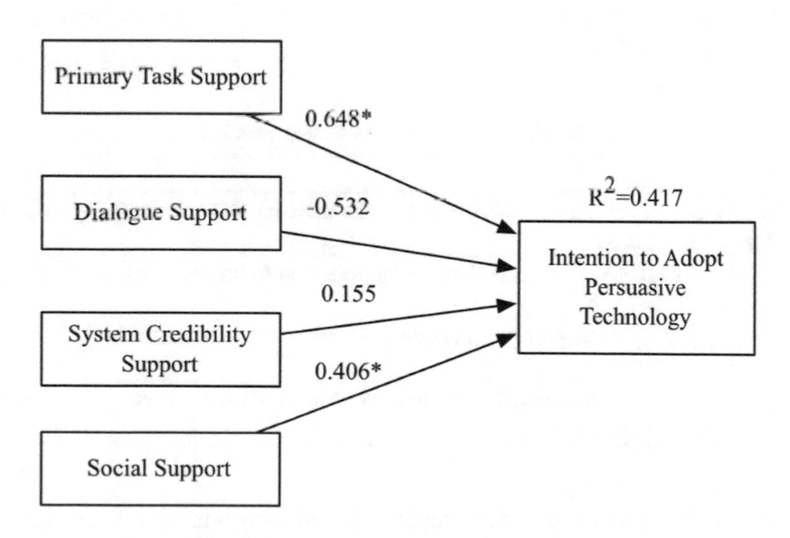

Fig. 2. The results of regression analysis.

5.3 Evaluating User Performance of Using Persuasive Technology

In the final stage of analysis, the study evaluates users' performance by comparing the test scores between current students with persuasive technology in class and previous students without persuasive technology in class. The result can estimate the impact of pedagogy with persuasive technology. The average scores of all classes presented in Table 4.

Table 4. Comparison of learning outcome.

Groups	N	$\overline{X}$	S.D.
Class with persuasive technology (second semester)	62	15.57	4.90
Class without persuasive technology 1 (first semester)	14	13.14	4.88
Class without persuasive technology 2 (first semester)	29	11.83	4.37
Class without persuasive technology 3 (first semester)	34	12.74	4.06

The result shows that the average scores of the students from class with applying persuasive technology in the second semester ($\overline{X} = 15.57$) were higher than the average scores from class without persuasive technology in the first semester ($\overline{X} = 13.14$, 11.83, and 12.74 respectively) at a significant level of 0.05.

6 Discussion and Conclusion

6.1 Hypotheses Testing Results

According to the results from the previous section, the hypotheses testing of this study can be concluded as (Table 5):

Table 5. Hypothesizes testing results

Hypotheses	Accepted
H1: Primary task support has a positive effect on the intention to use persuasive technology in the classroom	Yes
H2: Dialogue support has a positive effect on the intention to use persuasive technology in the classroom	No
H3: System credibility has a positive effect on the intention to use persuasive technology in the classroom	No
H4: Social support has a positive effect on the intention to use persuasive technology in the classroom	Yes

Primary task support and social support are two categories that have significantly positively affected with the intention to use persuasive technology in the classroom.

6.2 Discussion

This research proposed the framework for designing and evaluating persuasive technology in education. The research continues with the experiment. Firstly, the requirements on the persuasive technology in the education of participants were collected using a questionnaire. The result shows that the top ten of students' requirement on persuasive technology principles is (1) Social Learning (2) Social Facilitating (3) Tailoring (4) Suggestion (5) Normative Influence (6) Self-Monitoring (7) Social Role (8) Tunneling (9) Trustworthiness (10) Attractiveness.

Secondly, the study tried to find the factors affecting the intention to use persuasive technology in education. The most obvious finding to emerge from the analysis is that Task Support and Social Support are the only two factors that have significantly affected the intention to use persuasive technology in education. It can be explained that students will have a stronger intention to use computing technology in the classroom as tools and social actors. Hence, the lecturer, content provider, the application developer should consider those two factors in order to design the persuasive technology for education.

Lastly, the performance of the participants was evaluated by comparing the examination scores. The results showed that students who used persuasive technology in the classroom have a higher average score than the student who did not. Sine this difference has been found, it is probably due to the pedagogy that consists of persuasive technology that can be influence users' attitudes and behavior to pay attention in class. Therefore, the selected elements of persuasive technology should be keys to implementation in computing technology in education. A possible explanation for this might be that the participants in this study are the students aged between 18–24 years (Generation Z). This generation is the most active online user in Thailand [21] which means they are already adopted computer technologies. Previous research also suggested that this generation requires using the computing technology that integrates with social actors such as rewarding with positive feedback, providing social support, and observing other's performance [10]. The most accurate example of that system is the online social networks. It included the elements of persuasive technology for education as they have become the communication portal, which they attempt to light on an information-sharing activity conducted via online discussion [22].

In conclusion, in order to increase the efficiency and performance of learning, it is important for stakeholders in education to understand the principles of persuasive technology, and how to apply them in education before launching any pedagogy. Hence, the results of this study could be used as a guideline for the lecturer, content editor, learning facilitator and other stakeholders in education.

Finally, as the persuasive technology was able to improve students' outcomes. The framework for designing and evaluating persuasive technology in education could be able to apply with different levels of the students. Hence, future research should focus on the specific target group of students. Furthermore, the attractiveness of the user interface also should be considered as future research. This is in agreement with Ali et al.'s [23] finding which showed that mobile learning applications must be attractive. Since user interface plays the most important role for interaction between users and the application. Therefore, when conducting or choosing the persuasive technology for the next step, these principles should be concerned.

References

1. Tulgan, B.: Meet Generation Z: the second generation within the giant "Millennial" cohort, pp. 1–13. Rainmaker Thinking, Inc. (2013)
2. Turner, A.: Generation Z: technology and social interest. J. Individ. Psychol. **71**, 103–113 (2015)
3. Mohr, K.A.J., Mohr, E.S.: Understanding Generation Z students to promote a contemporary learning environment. J. Empower. Teach. Excell. **1**, 84–94 (2017)
4. Seemiller, C., Grace, M.: Generation Z goes to college. Plan. High. Educ. J. **44**(4), 105–107 (2016)
5. Malat, L., Vostok, T., Eveland, A.: Getting to Know Gen Z. Barnes & Noble College (2015)
6. Hallowell, E.M., Ratey, J.J.: Driven to distraction (revised): recognizing and coping with attention deficit disorder. Anchor (2011)
7. Rickes, P.C.: Generations in flux: how gen Z will continue to transform higher education space. Plan. High. Educ. **44**, 21–45 (2016)
8. Selwyn, N.: Making sense of young people, education and digital technology: the role of sociological theory. Oxford Rev. Educ. **38**(1), 81–96 (2012)
9. Mintz, J., Aagaard, M.: The application of persuasive technology to educational settings. Educ. Technol. Res. Dev. **60**(3), 483–499 (2012)
10. Fogg, B.J.: Persuasive Technology: Using Computers to Change What We Think and Do. Morgan Kaufman Publishers, Burlington (2002)
11. Fogg, B.J.: Creating persuasive technologies: an eight-step design process. In: Proceedings of the 4th International Conference on Persuasive Technology, Claremont, USA, pp. 1–6. ACM Digital Library (2009)
12. Oinas-Kukkonen, H., Harjumaa, M.: A systematic framework for designing and evaluating persuasive systems. Lecture Notes in Computer Science (Including Subseries Lecture Notes in Artificial Intelligence and Lecture Notes in Bioinformatics), vol. 5033, pp. 164–176. Springer, Heidelberg (2008)
13. Edwards, H.M., McDonald, S., Zhao, T., Humphries, L.: Design requirements for persuasive technologies to motivate physical activity in adolescents: a field study. Behav. Inf. Technol. **33**(9), 968–986 (2014)
14. Kadomura, A., Li, C.-Y., Tsukada, K., Chu, H.-H., Siio, I.: Persuasive technology to improve eating behavior using a sensor-embedded fork. In: Proceedings of the 2014 ACM International Joint Conference on Pervasive and Ubiquitous Computing, pp. 319–329 (2014)
15. Alhammad, M.M., Gulliver, S.R.: Online persuasion for e-commerce websites. In: The 28th International BCS Human Computer Interaction Conference on HCI, pp. 264–269 (2014)
16. Behringer, R., Soosay, M., Gram-Hansen, S., Øhrstrøm, P., Sørensen, C., Smith, C., Mikulecká, J., Winther-Nielsen, N., Winther-Nielsen, M., Herber, E.: Persuasive technology for learning and teaching – the EuroPLOT project. In: Proceedings of the International Workshop on EuroPLOT Persuasive Technology for Learning, Education and Teaching, pp. 3–7 (2013)
17. Goh, T.T., Seet, B.C., Chen, N.S.: The impact of persuasive SMS on students' self-regulated learning. Br. J. Educ. Technol. **43**, 624–640 (2012)
18. Filippou, J., Cheong, C., Cheong, F.: Designing persuasive systems to influence learning: modelling the impact of study habits on academic performance. In: PACIS 2015, pp. 156–169 (2015)
19. Dewi, Y., Widyasari, L., Nugroho, L.E., Permanasari, A.E.: Persuasive technology for enhanced learning behavior in higher education. Int. J. Educ. Technol. High. Educ. **16**(1), 15 (2019)

20. Hair, J.F., Black, W.C., Babin, B.J., Anderson, R.E.: Multivariate Data Analysis. Pearson-Hall, Upper Saddle River (2010)
21. We Are Social, Hootsuite: Digital in 2017: Southeast Asia. https://wearesocial.com/special-reports/digital-southeast-asia-2017. Accessed 10 Oct 2017
22. Susilo, A.: Exploring Facebook and Whatsapp as supporting social network applications for english learning in higher education. Teaching and Learning in the 21st Challenges for Lecturers and Teachers, pp. 10–24 (2014)
23. Ali, A., Alrasheedi, M., Ouda, A., Capretz, L.F.: A study of the interface usability issues of mobile learning applications for smart phones from the user's perspective. Int. J. Integr. Technol. Educ. **3**(1), 1–16 (2014)

Effectiveness of Six Text Classifiers for Predicting SET Stock Price Direction

Ponrudee Netisopakul(✉) and Woranun Saewong

Knowledge Management and Knowledge Engineering Laboratory,
Faculty of Information Technology, King Mongkut's Institute of Technology,
Ladkrabang, Bangkok, Thailand
Ponrudee@it.kmitl.ac.th, 60606080@kmilt.ac.th

Abstract. Six text classification methods were compared to find the best model for predicting Stock Exchange of Thailand stock prices. News headlines, on individual stocks, were classified as causing "change" and "no-change" based on a preset change threshold, 2.5%. The training dataset was collected by matching stock news in 2018 with stock names and filling in stock price changes. 258 news were associated with a "change" and 636 news with "no-change". The Thai text news items were preprocessed and converted to TF-IDF vector representation. Six machine learning text classification methods are applied to create six text classifier models and create a confusion matrix, then compared with actual changes to obtain accuracy scores. We found that a deep learning classifier (with 85.6% accuracy) scored better than other classifiers for one day price movement to assist short-term investments.

Keywords: Stock trends · Decision tree · Neural network · Deep learning · Naive Bayes · Random forest · SVM · Stock news · Text classification

1 Introduction

Although stock trading is an investment associated with high risk and high returns, the number of investors in Stock Exchange of Thailand (SET) doubled in four years from 1.3 million in 2013 to 2.5 million in 2017 [1]. Each year there are many new traders, without enough investment knowledge, and most of them often lose money. Therefore, the stock news, provided by many financial experts, can be important factors, affecting the stock market and each stock price movement. According to Cheng, more than 90% of investors use the news in stock trading decisions [2]. However, much stock news is difficult to interpret. This study filters only stock news and classifies it, so that investors can focus on probability that the news has a high chance of significantly moving the stock price.

Text mining extracts data from large textual datasets, consisting of structured, unstructured and semi-structured data and then turns this data into useful knowledge. We found that the text mining can be applied to filter finance news. Many researchers want to build a tool and predict the direction of the stock price [2–4] using text mining. This research determined the best model from six text classifiers: Decision Trees, Random Forests, Naïve Bayes, Support Vector Machine, Neural Networks and Deep

P. Meesad and S. Sodsee (Eds.): IC^{2}IT 2020, AISC 1149, pp. 104–118, 2020.
https://doi.org/10.1007/978-3-030-44044-2_11

Learning, and created an application to select Thai stock news that investors should study.

This paper is organized as follows: in Sect. 2, we review the background of text classification, the various methods and other related research on analyzing share prices on foreign stock exchanges. In Sect. 3, we describe the methodology used here, and in Sect. 4, we present the results and discuss them. In Sect. 5, we conclude and suggest a plan for future work.

2 Related Work

2.1 Predictive Data Mining Algorithms

A prediction is the ultimate goal of most data mining algorithms. A predictive model can be created, using machine learning from training data, for example, a classification, a regression or a text mining algorithm. Predictive data mining usually applies many algorithms to create many models, appropriate for the problem on hand. Considerable previous research has applied text mining from stock news for predicting stock changes: for example, Cheng [2] used Taiwan stock market news and support vector machine classifiers to forecast intraday stock price changes. Ichinose and Shimada [3] used data from Yahoo Japan finance news to predict the Nikkei stock using a support vector machine and a linear perceptron. Kumar et al. [5] used multiple classifiers to find the most accurate classifier. Each paper forecasts intraday stock price changes to find the best method for predicting price change trends from finance news. We extended existing research with experiments using six text classifiers applied to Thai stock news. The six classifiers are described as follows.

Decision Tree
A decision tree model is a hierarchy of segmented data, using the information gained in the contribution of each attribute. The root node is the attribute with the largest information gain, the subsequent child nodes are associated with the next level of the largest information gain. Subsequently, each branch is labeled with the outcome of a condition on the segmented data. Finally, each leaf node holds a class label.

Support Vector Machine
A Support Vector Machine (SVM) is a discriminative classifier, formally defined by a separating hyperplane. This machine performs regression and classification tasks by constructing nonlinear decision boundaries using some kernel functions. This suits some types of feature spaces, in which these boundaries are found. An SVM can exhibit a large degree of flexibility in handling classification and regression tasks of varied complexities. The common kernel functions of SVM models can be either linear, polynomial, radial basis functions (RBFs) or sigmoid functions.

Random Forest
A random forest algorithm is a supervised classification algorithm that creates a forest with many decision trees. Each tree predicts a class and the ensemble algorithm collects the majority vote of all trees. Figure 1 illustrates this process.

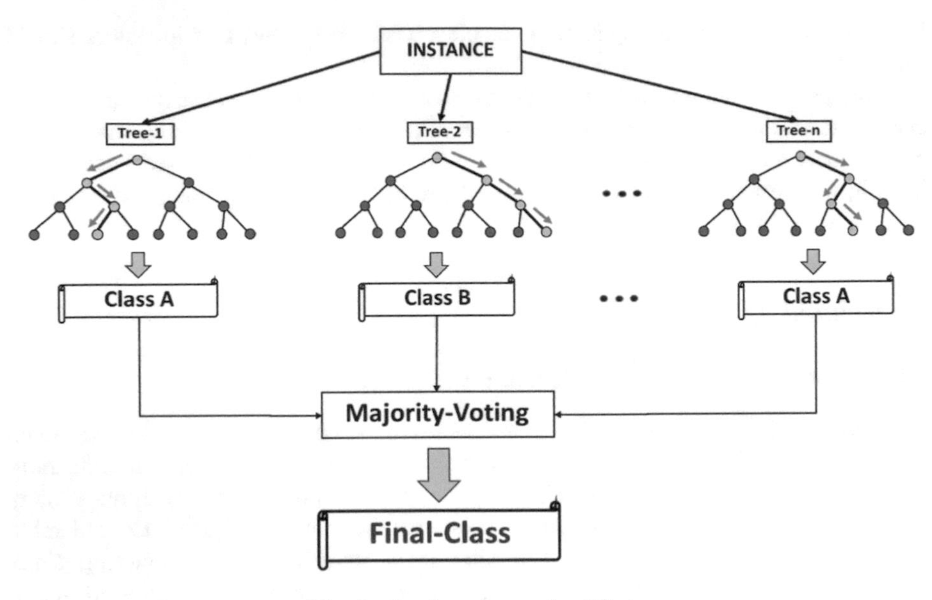

Fig. 1. Random forest simplified.

Naïve Bayes

In machine learning, a naïve Bayes classifier [6] is a form of supervised learning. A naïve Bayes classifier is based on principles of conditional probability, from Bayes' theorem, where the conditional probability of an event, Y, given another event, X, that has occurred, is calculated. Bayes' theorem calculates the posterior probability, P(Y|X), from P(Y), P(X), and P(X|Y):

$$P(Y|X) = \frac{P(Y)P(X|Y)}{P(X)} \tag{1}$$

In the Sklearn Python Library [7], a multinomial model is the event model for working with text classification. The advantage of the multinomial naïve Bayes classifier is that it is easy to implement, highly scalable, with a number of predictors and data points, and also works with term frequency–inverse document frequency (tf-idf) vectorization [8].

Artificial Neural Network

An Artificial Neural Network (ANN) is a machine learning framework that attempts to mimic the function of human neurons. The simplest neural network consists of only one neuron, called a perceptron. Multiple inputs, modified by a weight function, can be obtained and passed into an activation function to produce an output. Multi-layer perceptrons are ANNs, which consist of one input layer, one or more hidden layers, and one output layer. For a feedforward ANN, each node in one layer is connected to nodes, in the next layer, with different weights. The weighted sum of all previous layer nodes is then calculated for each node, in the next layer, and passed through an activation function to produce an output from that node contributing to nodes in the next layer, until they are all reach the output layer. The weights can then be adjusted in a backward fashion, until the ANN reaches the optimal state.

Deep Learning
Deep learning methods or deep neural networks are ANNs with two or more hidden layers. In theory, different layers should be responsible for processing different sets of features, based on previous layer output. Many types of activation functions are used in deep neural networks. Those include step, sigmoid, ReLU and tanh functions.

2.2 Performance Evaluation

To evaluate the performance of each classification model, we used confusion matrix [9]. The confusion matrix can be represented as a 22 matrix, as shown in Fig. 2. The formulae for accuracy, precision, recall and f1 score as shown in Eqs. (2), (3), (4).

Prediction \ Actual	Change	No Change
Change	**True Positive (TP)**	**False Positive (FP), Type I Error**
No Change	**False Negative (FN), Type II Error**	**True Negative (TN)**

Fig. 2. Confusion matrix.

Accuracy
The accuracy measures how often the classifier was correct, in both change and no change classes.

$$Accuracy = \frac{(TP + TN)}{Total\ Population} \tag{2}$$

Precision
The precision measures the ratio of correct positive instance prediction (in change class) per all positive instance prediction.

$$Precision = \frac{TP}{(TP + FP)} \tag{3}$$

Recall
The recall measures the ratio of correct positive instance prediction (in change class) per true positive instance.

$$Recall = \frac{TP}{(TP + FN)} \tag{4}$$

F1 Score

The F1 score was the harmonic average of the precision and recall. The best score was 1 and the worst score was zero.

$$F1 = \frac{(2 \times Recall \times Precision)}{(Recall + Precision)} \tag{5}$$

3 Methodology

We aimed to find the best classification model for filtering stock news associated with high chances of price changes, i.e. news that investors should read carefully. Firstly, we explain the definitions and assumptions in this research.

3.1 Assumptions

- Investors should pay attention to stock news containing one or more stock names and affecting the price of the associated stock on the same day. That is, the closing stock price must be significantly higher or lower than the previous closing price.
- A threshold percentage, such as 2 to 5%, can be preset to reflect significant changes. Percent price change calculated from current and previous closing prices.

$$Change = \frac{current - previous}{previous} \times 100 \tag{6}$$

  Changes higher than the preset threshold are labeled "Surge", and lower than the threshold are labeled "Plunge". The area between Surge and Plunge is labeled "Stable".
- To simplify the classifier task, we only classified stock news into two classes: "change" and "no-change". The "change" class combines data from "Surge" and "Plunge" classes, to which investors should pay attention. The "no-change" class contains data where the prices are "Stable".

In the next paragraph, we explain our research methodology. The designed system was divided into two phases as shown in Fig. 4. The first phase was text preprocessing and tf-idf vectorization [8] where the input was stock news retrieved and extracted from a financial news webpage and stored in a dataset. Thai text news in the dataset passes through various preprocessing steps. The output is a vector of phrases prepared for the second phase. The second phase was the training and evaluating classifiers. Some data was used as pre-training data. Pre-training obtained optimal values of parameters for some classifiers, such as SVMs and ANNs. The outputs of the training process were predicted classifiers. A confusion matrix was derived by comparing predicted and actual classes. Each of the classifiers was evaluated to find the model, using the confusion matrix which best calculated precision, recall, f1 score and accuracy.

3.2 Text Processing

This step obtained stock news from a financial news webpage [10] and daily stock prices from the SET market [11]. Stock symbols were extracted from the news and matched to stock prices on the current and previous day. The label "change" or "no-change" was attached to each news item based on the price change explained before. Finally, we brought the news and the label to the dataset using PHP language as shown in Fig. 3. We will explain details of all the processes as follows.

Data Collection
First, stock news were collected using PHP on the website: www.hoonsmart.com [10] from January to December, 2018 – a total of 6,349 items. Each news item consisted of headlines news, news, stock symbol, date, time and URL. In the headline and body of the news item, we removed noise words, for example, multiple white spaces, tabs and punctuation characters. Daily stock prices were also collected from the website www.setsmart.com [11]. Only current and previous day closing prices were collected. The output of this stage was a dataset of all of the stock news articles and previous closing prices.

Stock News Filtering
Stock news was manually filtered using the following criteria.

- The news tag does not contain a company or security in the Thai stock market. For example, "ตลาดหุ้นโตเกียวเปิดร่วง – นักลงทุนขายทำกำไร" (translated as "Tokyo stocks fell at the open – investors sell to make profit").
- The news and headlines were general news and did not contain a company or security in the Thai stock market. For example, "รังสิตโพลล์ครั้ง 5 ประยุทธ์ จันทร์โอชา อันดับ 1" (translated as "Rungsit poll number 5 Prayoot ChanOcha number 1").
- Article was written by columnist or analysts. The article was analyzed and often biased by personal opinions or inclinations rather than facts. Therefore, we decided to filter them out. For example, "คอลัมน์เสือใหญ่: บิ๊กเคส 2561 พระเอกตัวจริง" (translated as "Seuyai column: Big Case 2561, the real hero").

If an article contained more than one stock symbol, it was duplicated and labeled with each stock symbol. Our aim was to find the best model for predicting the direction of a stock price after the news release. To speed up the processing time, only headline news, date, and the stock symbol were selected for further processing. After filtering, 1,006 articles remained.

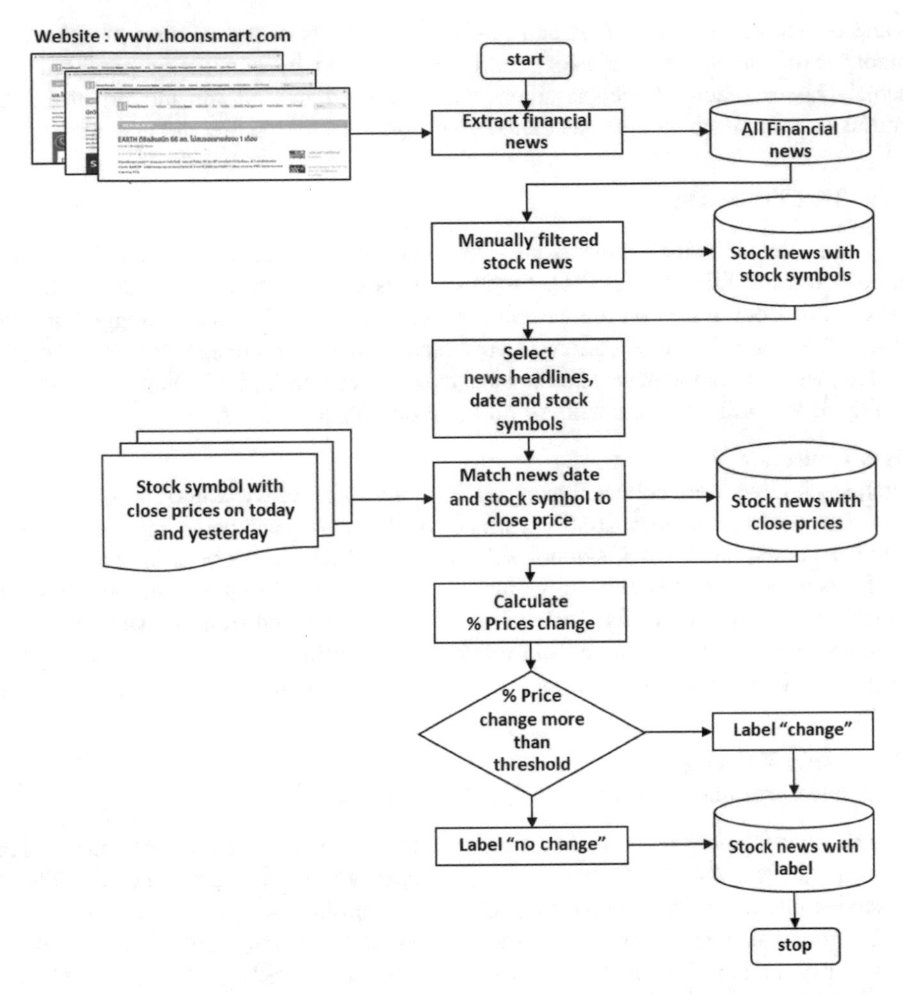

Fig. 3. Text processing steps to create a database of articles with classification labels.

Matching News and Closing Price

There were 357 companies in the stock news collected from January to December 2018. Date formats in the stock news and the closing price sources were different and they were converted to a common format. After that the article day and the stock symbol from the headline were matched to the stock price on the same date. So, the new dataset consisted of headlines news, closing prices on the current and previous day. 894 articles remained, because some headlines mentioned IPO stocks and appeared before any price data was available. In addition, headlines appearing on Saturday or Sunday, without trading days and matching prices, were excluded from the dataset.

Calculating Price Change and Adding a Label

In this stage, we calculated the price change (as a percent of previous close) using Eq. (5). From preliminary experiments, we found that the threshold of ±2.5% price change was associated with significant effects.

$$\text{"change"}\ class = price\ change\ \geq |2.5\%| \tag{7}$$

$$\text{"no-change"}\ class = price\ change\ < |2.5\%| \tag{8}$$

In the “change” class, with both of “surge” and “plunge” labels 258 items were identified, leaving 636 in the “no-change” class. The output of this stage had 894 items. Each row of this dataset consists of headline and class. For example, “7UP ขายโรงไฟฟ้าชีวภาพมูลค่า 102 ล้าน-ซื้อกิจการ ICT” (translated as 7UP sold bio power plants at 102 million and bought ICT business, “no-change”).

Word Segmentation Using Deepcut

The following steps were developed using Python. First, from each sentence, a list of words was created using the deepcut library [12]. For example, [‘AJA’, ‘ฮึด’, ‘สู้’, ‘ดีน’, ‘หนี’, ‘ฟลอร์’, ‘ที’, ‘สอง’]. The other output from this step was a bag of words, containing all the words from any sentence.

Training Set for Five-Cross Validation

Because the dataset was not very large and also unbalanced, we created a training set using random sampling with replacement, that is, random sampling 500 “change” headlines from 258 items in “change” training class and 500 “no-change” headlines from 636 items in “no-change” training class. Then, we divided it into five equal parts: four parts were used for training and one part was used for testing so that each part was in the testing set for one round and in the training set for four rounds, i.e. a five-fold cross validation. Therefore, each fold contained 100 headlines from the “change” class and 100 from the “no-change” class. This method led to better classification results than using the raw dataset.

TF-IDF Vectorization

As the computer was unable to calculate the text, we converted the output segmented words, using term frequency – inverse document frequency (tf-idf) vectorization [8], to numeric vectors using the following formula.

$$tf - idf(t) = tf(t) \times idf(t) \tag{9}$$

Where

$$tf(t) = \frac{Number\ of\ times\ term\ t\ appears\ in\ a\ document}{Total\ number\ of\ terms\ in\ the\ document} \tag{10}$$

This function can find the importance of the word in the document at the weight of each word.

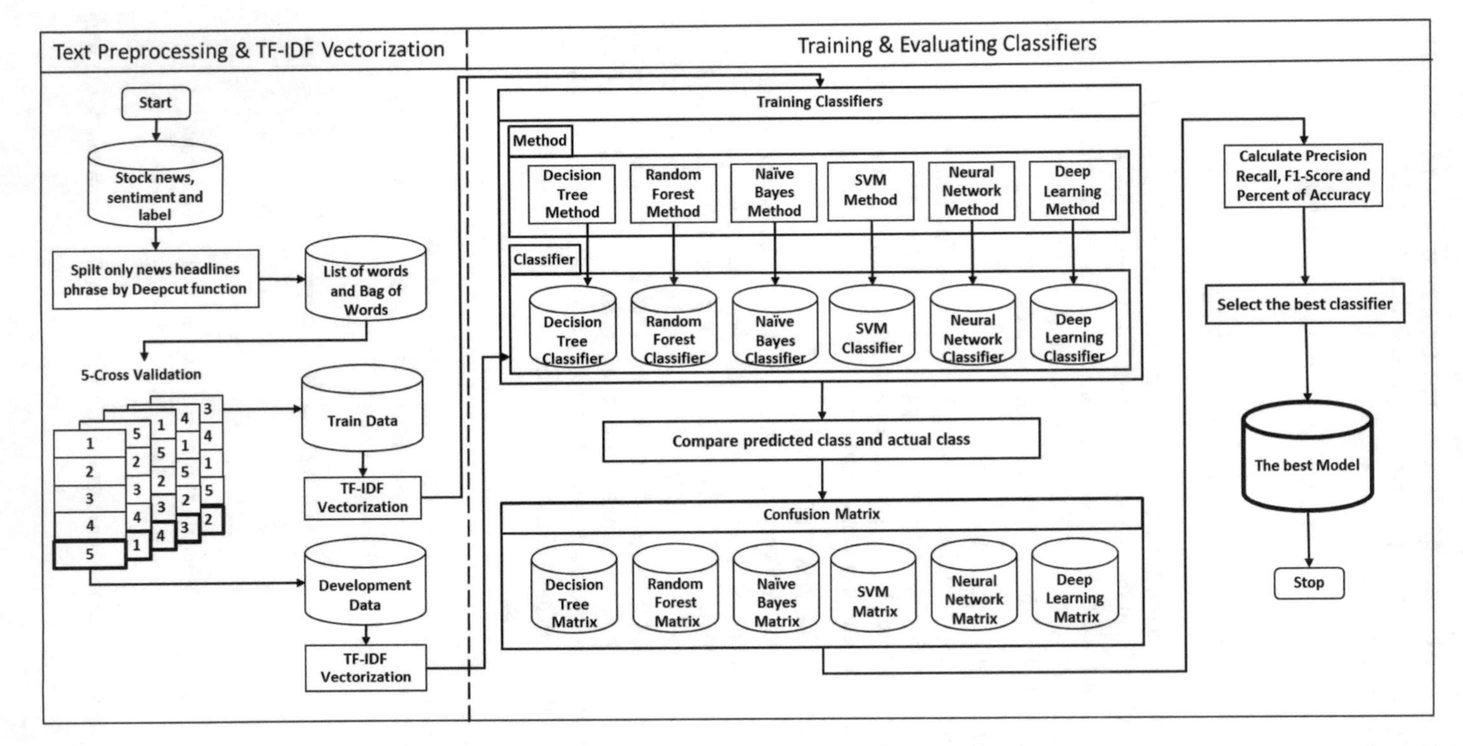

Fig. 4. System designed to find the best classification model for filtering stock news.

Classification Models

A classification model or a classifier was a result of the classification training method. Each classifier was established using a tf-idf vector representation of stock news and their classes was as a training set. Before training, the parameters fit each method were obtained during pre-training. Pre-training used a subset of 400 balanced class headlines.

We compared six classification methods: Decision Trees, Random Forests, Naïve Bayes, Support Vector Machines, Neural Networks and a Deep Learning Classifier. Each method led to different classifiers, which were then used to predict the class of the test dataset as either "change" or "no-change". The resulting confusion matrix is discussed in Sect. 4 Result and Evaluation.

Evaluating Classifiers

The predicted and actual classes from the classifiers were compared and shown as a confusion matrix. The precision, recall, f1 and accuracy score of each classifier were calculated from the confusion matrix. Finally, we compared accuracy scores to find the best model, as shown in the next section.

4 Result and Evaluation

Here, we explain the parameters in each model from pre-training process and the resulting confusion matrix.

4.1 Effectiveness of Each Classifier

Deep Learning

The parameters of deep learning model were the activation function, which controls the amplitude of the output and the number of neurons and layers. We used a three-layer system. In the first layer, the activation function was set to "ReLU" and the neurons in the input layer was set to 128. In the second layer, the activation function was set to "ReLU" and there were 10 neurons in the hidden layer. In the last layer, the activation function was set to "sigmoid" and there was only one neuron in the output layer. In addition, we used backpropagation algorithm for adjusted a weights by error in the network on each neurons. In backpropagation algorithm, the optimizer is set to "adam" and loss is set to "binary_crossentropy".

Figure 5 shows results from the deep learning model. Overall accuracy was 85.6%.

Fold No.	TP	FP	TN	FN	Change			No-Change			Accuracy
					Precision	Recall	F1	Precision	Recall	F1	
1	91	9	78	22	0.805	0.910	0.854	0.897	0.780	0.834	84.5%
2	79	22	94	6	0.929	0.782	0.849	0.810	0.940	0.870	86.1%
3	90	10	84	16	0.849	0.900	0.874	0.894	0.840	0.866	87.0%
4	93	7	86	14	0.869	0.930	0.899	0.925	0.860	0.891	89.5%
5	90	10	72	28	0.763	0.900	0.826	0.878	0.720	0.791	81.0%
Avg	**88.60**	**11.60**	**82.80**	**17.20**	**0.843**	**0.884**	**0.860**	**0.881**	**0.828**	**0.851**	**85.6%**

Fig. 5. Confusion matrix from deep learning model.

Random Forest

The parameters of the random forest model were splitting criterion and max_depth. The splitting criterion was set to "entropy" and the max_depth was set to 35.

As shown in Fig. 6, overall, the random forest model had and accuracy of 85.3%.

Fold No.	TP	FP	TN	FN	Change			No-Change			Accuracy
					Precision	Recall	F1	Precision	Recall	F1	
1	81	19	84	16	0.835	0.810	0.822	0.816	0.840	0.828	82.5%
2	79	21	90	10	0.888	0.790	0.836	0.811	0.900	0.853	84.5%
3	89	11	86	14	0.864	0.890	0.877	0.887	0.860	0.873	87.5%
4	90	10	86	14	0.865	0.900	0.882	0.896	0.860	0.878	88.0%
5	87	13	81	19	0.821	0.870	0.845	0.862	0.810	0.835	84.0%
Avg	**85.20**	**14.80**	**85.40**	**14.60**	**0.855**	**0.852**	**0.852**	**0.854**	**0.854**	**0.853**	**85.3%**

Fig. 6. Confusion matrix from random forest model.

Decision Tree

The parameters of the decision tree model were splitting criterion and max_depth. The splitting criterion was set to "gini" and max_depth was set to 40.

Figure 7 shows results using decision trees. Overall accuracy was 84.6%.

Fold No.	TP	FP	TN	FN	Change			No-Change			Accuracy
					Precision	Recall	F1	Precision	Recall	F1	
1	89	11	79	21	0.809	0.890	0.848	0.878	0.790	0.832	84.0%
2	80	20	83	17	0.825	0.800	0.812	0.806	0.830	0.818	81.5%
3	93	7	78	22	0.809	0.930	0.865	0.918	0.780	0.843	85.5%
4	92	8	83	17	0.844	0.920	0.880	0.912	0.830	0.869	87.5%
5	91	9	78	22	0.805	0.910	0.854	0.897	0.780	0.834	84.5%
Avg	**89.00**	**11.00**	**80.20**	**19.80**	**0.818**	**0.890**	**0.852**	**0.882**	**0.802**	**0.839**	**84.6%**

Fig. 7. Confusion matrix of decision tree model.

Support Vector Machine

The parameters of SVM model were the kernel type and penalty parameter C. The kernel type was set to "linear" and penalty parameter C was set to 1.

Figure 8 shows results using a SVM. Overall accuracy was 84.1%.

Fold No.	TP	FP	TN	FN	Change			No-Change			Accuracy
					Precision	Recall	F1	Precision	Recall	F1	
1	84	16	80	20	0.808	0.840	0.824	0.833	0.800	0.816	82.0%
2	82	18	86	14	0.854	0.820	0.837	0.827	0.860	0.843	84.0%
3	89	11	83	17	0.840	0.890	0.864	0.883	0.830	0.856	86.0%
4	93	7	83	17	0.845	0.930	0.886	0.922	0.830	0.874	88.0%
5	85	15	76	24	0.780	0.850	0.813	0.835	0.760	0.796	80.5%
Avg	**86.60**	**13.40**	**81.60**	**18.40**	**0.825**	**0.866**	**0.845**	**0.860**	**0.816**	**0.837**	**84.1%**

Fig. 8. Confusion matrix of SVM model.

Neural Network

The parameters of the neural network model were the same as the deep learning model. However, the neural network was set to 2 layers. For the input layer, the activation function was "ReLU" and there were 500 neurons. For the output layer, the activation function was "sigmoid" and there was a single neuron. We used backpropagation algorithm, the optimizer is set to "adam" and loss is set to "binary_crossentropy".

As shown in Fig. 9, overall accuracy for the neural network model was 81.1%.

Fold No.	TP	FP	TN	FN	Change			No-Change			Accuracy
					Precision	Recall	F1	Precision	Recall	F1	
1	89	11	65	35	0.718	0.890	0.795	0.855	0.650	0.739	77.0%
2	79	21	86	14	0.849	0.790	0.819	0.804	0.860	0.831	82.5%
3	95	5	69	31	0.754	0.950	0.841	0.932	0.690	0.793	82.0%
4	96	4	73	27	0.780	0.960	0.861	0.948	0.730	0.825	84.5%
5	92	8	67	33	0.736	0.920	0.818	0.893	0.670	0.766	79.5%
Avg	**90.20**	**9.80**	**72.00**	**28.00**	**0.768**	**0.902**	**0.827**	**0.887**	**0.720**	**0.791**	**81.1%**

Fig. 9. The confusion matrix of neural network model.

Naïve Bayes

As shown in Fig. 10, the naïve Bayes model had an average overall accuracy of 80.2%.

Fold No.	TP	FP	TN	FN	Change			No-Change			Accuracy
					Precision	Recall	F1	Precision	Recall	F1	
1	86	14	74	26	0.768	0.860	0.811	0.841	0.740	0.787	80.0%
2	80	20	85	15	0.842	0.800	0.821	0.810	0.850	0.829	82.5%
3	92	8	73	27	0.773	0.920	0.840	0.901	0.730	0.807	82.5%
4	91	9	80	20	0.820	0.910	0.863	0.899	0.800	0.847	85.5%
5	82	18	29	29	0.739	0.820	0.777	0.617	0.500	0.552	70.3%
Avg	**86.20**	**13.80**	**68.20**	**23.40**	**0.788**	**0.862**	**0.822**	**0.814**	**0.724**	**0.764**	**80.2%**

Fig. 10. Confusion matrix of naïve bayes model.

4.2 Comparing the Six Classifiers

The six classifiers are compared in Fig. 11. The highest and the second highest accuracy models were the deep learning classifier (85.6%) and the random forest classifier (85.3%). However, accuracies for all models were above 80%.

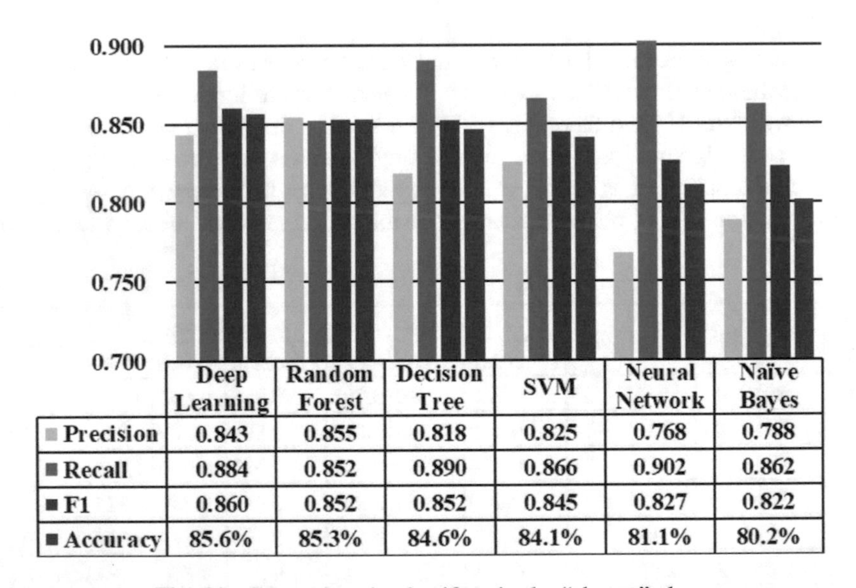

Fig. 11. Comparing six classifiers in the “change” class.

5 Conclusion

We extracted stock news from two online websites. After cleaning and combining data from the two sources, headline items were labeled as indicating “change” or “no-change”, where a percentage price change threshold determined whether an item will affect price or not. In data collected from January 1 to December 31, 2018, after cleaning and preprocessing, 894 headline items remained with 258 items labeled as "change" and 636 as "no-change". Deepcut tools and tf-idf vectorization were applied to covert Thai text headlines to number vectors.

Six classifiers - decision trees, deep learning, random forests, naïve Bayes, neural networks and SVM - were assessed. A pre-training data set of 400 items was used to fit parameters for each classifier. A five-fold cross validation approach compared the accuracies of each method. Each fold contained 100 "change" headline items and 100 "no-change" headline items. Similarly in the training dataset, the five-fold cross validation contained 500 "change" class headlines and 500 "no-change" class headlines in the total of 1,000 headlines.

Evaluation showed that the best model was the deep learning classifier for an accuracy score of 85.6%. However, all the tested classifiers scored above 80%. This indicates that these classifiers can be used with reasonably accuracy to predict whether a news item will affect the stock price. However, since the dataset compared only the price change on the current and previous day, the result is only appropriate for short-term investments.

In future work, the assumption that the stock news only affects the stock price immediately should be relaxed, because the news may reflect the long-term price in a week, a month, or a year.

Acknowledgement. We would like to thank the Thai NLP Group for sharing their knowledge and resources, King Mongkut's Institute of Technology Ladkrabang (KMITL) for the research funding and KMITL KRIS for advice on technical English.

References

1. The Stock Exchange of Thailand: TSD's Statistical Highlights (As of December). https://www.set.or.th/tsd/en/download/statistic.html. Accessed 20 Jan 2019
2. Cheng, S.H.: Forecasting the change of intraday stock price by using text mining news of stock. In: Proceedings of the Ninth International Conference on Machine Learning and Cybernetics, pp. 2605–2609. IEEE, Qingdao (2010)
3. Ichinose, K., Shimada, K.: Stock market prediction from news on the web and a new evaluation approach in trading. In: Proceedings of 5th IIAI International Congress on Advanced Applied Informatics (IIAI-AAI), pp. 77–81. IEEE, Kumamoto (2016)
4. Yetis, Y., Kaplan, H., Jamshidi, M.: Stock market prediction by using artificial neural network. In: Proceedings of world Automation Congress (WAC), pp. 1–5. IEEE, Waikoloa, HI, USA (2014)
5. Kumar, I., Dogra, K., Utreja, K., Yadav, P.: A comparative study of supervised machine learning algorithm stock market trend prediction. In: Proceedings of 2018 Second International Conference on Inventive Communication and Computational Technologies (ICICCT), pp. 1003–1007. IEEE, Coimbatore (2018)
6. Loon, R.V.: Naive Bayes classifier with example simplilearn channel. https://youtu.be/l3dZ6ZNFjo0. Accessed 03 Aug 2019
7. Pedregosa, F., et al.: scikit-learn: sklearn.naive_bayes.MultinomialNB. https://scikit-learn.org/stable/modules/generated/sklearn.naive_bayes.MultinomialNB.html. Accessed 11 Aug 2019

8. Pedregosa, F., et al.: scikit-learn: learn.feature_extraction.text.TfidfVectorizer. https://scikit-learn.org/stable/modules/generated/sklearn.feature_extraction.text.TfidfVectorizer.html. Accessed 11 Aug 2019
9. Pedregosa, F., et al.: scikit-learn: machine learning in Python. https://scikit-learn.org/stable/modules/generated/sklearn.metrics.confusion_matrix.html. Accessed 20 July 2019
10. Chairattanamanokorn, N., et al.: HoonSmart Breaking News. https://www.hoonsmart.com. Accessed 20 Jan 2019
11. The Stock Exchange of Thailand.: Companies/Securities in Focus Historical Trading. https://www.setsmart.com. Accessed 05 May 2019
12. Viriyayudhakorn, K.: Thai Natural Language Processing (Thai NLP) Resource. https://github.com/kobkrit/nlp_thai_resources. Accessed 12 Aug 2019

Biomarker Identification in Colorectal Cancer Using Subnetwork Analysis with Feature Selection

Sivakorn Kozuevanich[1](✉), Asawin Meechai[1], and Jonathan H. Chan[2]

[1] Department of Chemical Engineering, King Mongkut's University of Technology Thonburi, Bangkok, Thailand
sivakorn.koz@gmail.com, asawin.mee@kmutt.ac.th
[2] School of Information Technology, King Mongkut's University of Technology Thonburi, Bangkok, Thailand
jonathan@sit.kmutt.ac.th

Abstract. Gene Sub-Network-based Feature Selection (GSNFS) is an efficient method for handling case-control and multiclass studies for gene sub-network biomarker identification by an integrated analysis of gene expression, gene-set and network data. However, GSNFS has produce considerably high number of sub-network and has not assessed the importance of each sub-network. Recently, we have incorporated 2 feature selection techniques; correlation-based and information gain into the GSNFS workflow to help reduce the number and assess the importance of each individual sub-network. The extended GSNFS method was clearly shown to identify good candidate gene subnetwork markers in lung cancer. In this work, we applied a similar work flow to colorectal cancer. First, the top- and bottom- 5 ranked gene-sets were selected and investigated the classification performance. Similarly, the top-ranked gene-sets showed a better performance than the bottom-ranked gene-sets. The identified top-ranked gene-sets such as TNF-beta and MAPK signaling pathway were known to relate to cancer. In addition, the characteristic of top identified pathway network was further analyzed and visualized. SMAD3, a gene that was reported to be related to cancer by many studies, was mostly found to have the highest neighbor in 4 datasets. The results in this study has confirmed that GSNFS combined with feature selection is very promising as significantly fewer subnetworks were needed to build a classifier and gave a comparable performance to a full dataset classifier.

Keywords: Gene expression analysis · Gene-set · Classification · Colorectal cancer · Correlation-based feature selection · Information gain feature selection

1 Introduction

Cancer is considered to be one of the most complex diseases that has complicated treatment process until now. An important challenge for researchers is to classify a patient to an appropriated cancer class and identify essential biomarkers. Biomarker is the most valuable component in the cancer research and can be beneficial to the treatment process. In addition, time and cost can be significantly reduced in the diagnosis of cancer on a

P. Meesad and S. Sodsee (Eds.): IC²IT 2020, AISC 1149, pp. 119–127, 2020.
https://doi.org/10.1007/978-3-030-44044-2_12

patient and most importantly, cancer can be coped in advance with the help of biomarkers. Thus, the identification of relevant biomarker is a primary task for a researcher [1]. Several traditional gene-based methods were used to identify biomarkers in the past. It was developed based on microarray data alone with an assumption that each gene contributes independently to clinical outcomes. Low reproducibility of the prediction performance was detected when tested across different datasets due to the noisiness and the molecular and cellular heterogeneity. This method does not effectively deal with high-dimensional data and poor biomarkers were identified [2]. Gene-set or pathway-based method were developed after gene-based method. This method integrates the gene-set or pathway data that are involved in disease or biological process and produces gene-set biomarkers. It uses the pathways membership information for a better understanding of the underlying biological mechanisms [3]. Network-based method uses interaction structure to identify sub-network biomarkers. Commonly used network data are protein-protein (PPI) and gene-gene interaction (GGI) [4]. In 2016, Doungpan et al. proposed Gene Sub-Network-Based Feature Selection (GSNFS) that incorporates gene expression, gene-set, and network data to identify lung cancer sub-network biomarkers [5]. Although this method has proved to offer gene subnetwork markers with good classification performance, considerably high numbers of gene sub-networks were identified. In 2011, Chan et al. [6] proposed the use of ranker feature selection method to rank and select pathway markers based on their activities. The results have showed that feature selection methods with top 3 pathway markers were suitable for both logistic and NB classifiers. Recently, our group [7] has improved GSNFS by applying feature selection techniques to reduce the number of sub-network biomarkers in a classifier model using lung cancer datasets. It has been shown that the top-ranked sub-networks provided higher performance than the bottom-ranked sub-networks. Some well-known cancer related pathways such as the MAPK signaling pathway has also been commonly found in the identified top-ranked gene-sets. Furthermore, combined top-ranked gene-sets from top 2 up to top 30 showed a further improvement on the performance when compared to using individual gene-sets. In addition, approximately 5–10 combined top gene-sets to build a classifier gave the highest overall performance and may be taken to be a suitable range of sub-network biomarkers.

Here, we aim to verify and further improve our previous studies [7] by applying same feature selection method on four colorectal cancer datasets and additionally explore the network characteristic of the top identified gene-set. We apply a similar work flow as we did on our previous studies. We firstly investigate the effect of top- and bottom-ranked sub-networks on the classification performance. The results are expected to show a similar trend as obtained from using lung cancer [7]. In addition, top identified gene-set network will be further analyzed to explore the characteristic of that gene-set network.

2 Materials and Methods

2.1 Materials

Datasets

Expression Data. Four colorectal cancer microarray datasets obtained from Gene Expression Omnibus or GEO database (http://www.ncbi.nih.gov/geo/) [8] were used as shown in Table 1. Furthermore, each microarray dataset was preprocessed by discarding the microarray probe representing multiple genes in order to avoid the ambiguous interpretation of those genes. Then the expression values were normalized by z-transformation into standard scores having mean of zero and a standard deviation of one for further analysis.

Table 1. Microarray gene expression datasets.

Dataset	Reference	Disease	No. of samples	
			Normal	Tumor
GSE4107	[10]	Colorectal cancer	10	12
GSE8671	[11]	Colorectal cancer	32	32
GSE9348	[12]	Colorectal cancer	70	12
GSE32323	[13]	Colorectal cancer	17	17

Gene-set Data. Gene-set data comprising 353 gene-sets were retrieved from PathwayAPI [9] which is a web service for the curated gene-set data to provide gene-gene relationship.

Network Data. Protein-protein interaction (PPI) data were obtained from the Biological General Repository for Interaction Datasets (BioGRID) (http://thebiogrid.org/) which was proposed by Stark et al. [14]. They were first pre-processed to obtain only the human gene and protein interactions. The PPI network composed of 16,041 genes/proteins and 213,996 interactions.

2.2 Methodology

In this study, we employed the similar work flow as used in our previous studies [7]. First, significant sub-networks (gene-sets) were identified by integration of microarray, gene-set, and network data in GSNFS. Formerly in GSNFS, sub-networks with lower than 4 gene members were not considered to be a sub-network and were discarded. Similar to our previous work, all identified sub-network regardless of number of gene member in the sub-network were considered to determine the outcome of different ranked sub-networks on the classification performance as some fewer gene members gene-sets might outperform those with higher member. Later, correlation-based and information gain evaluator in Weka tools [15] were applied to rank those identified gene-sets. Two different feature selection techniques were intentionally applied for comparison and confirmation purposes. Correlation-based and information gain evaluators were suitable choices for numeric attributes in our datasets. The attribute in this case represents each gene-set/pathway with its corresponding significant genes. Then the classification performance of using top and bottom 5 attributes were investigated. Also, the top-ranked identified gene-sets and its occurrence from using 2 attribute evaluators across 4 datasets were evaluated. Lastly, the characteristic of top identified gene-set network was analyzed and visualized by using Cytoscape [16].

3 Results and Discussion

In our study, ROC (receiver operating curve) area (area under the curve) was chosen to represent the classification performance of attributes. The classification performance of each individual top and bottom 5 rank attributes is shown in Fig. 1. The black bar represents the results on the 10-fold cross validation of training datasets. The 3 other bars were results from testing the model with different datasets. Outstanding performance of the top-ranked gene-sets can be observed from training datasets. On the part of testing the model across different datasets, top-ranked gene-sets averagely has a better performance than the bottom-ranked gene-sets. Similarly, the performance from using information gain attribute evaluator gave an analogous result as shown in Fig. 2. Dominant performance of top-ranked gene-sets can be observed from training datasets. The top-ranked gene-sets averagely gave a higher performance than those in the bottom-ranked gene-sets when tested with different datasets. Furthermore, by comparing the results of the top-ranked gene-sets to using all 353 gene-sets as a classifier model as shown in Table 2, most of the individual top-ranked gene-sets showed a comparable performance with significantly lower data usage.

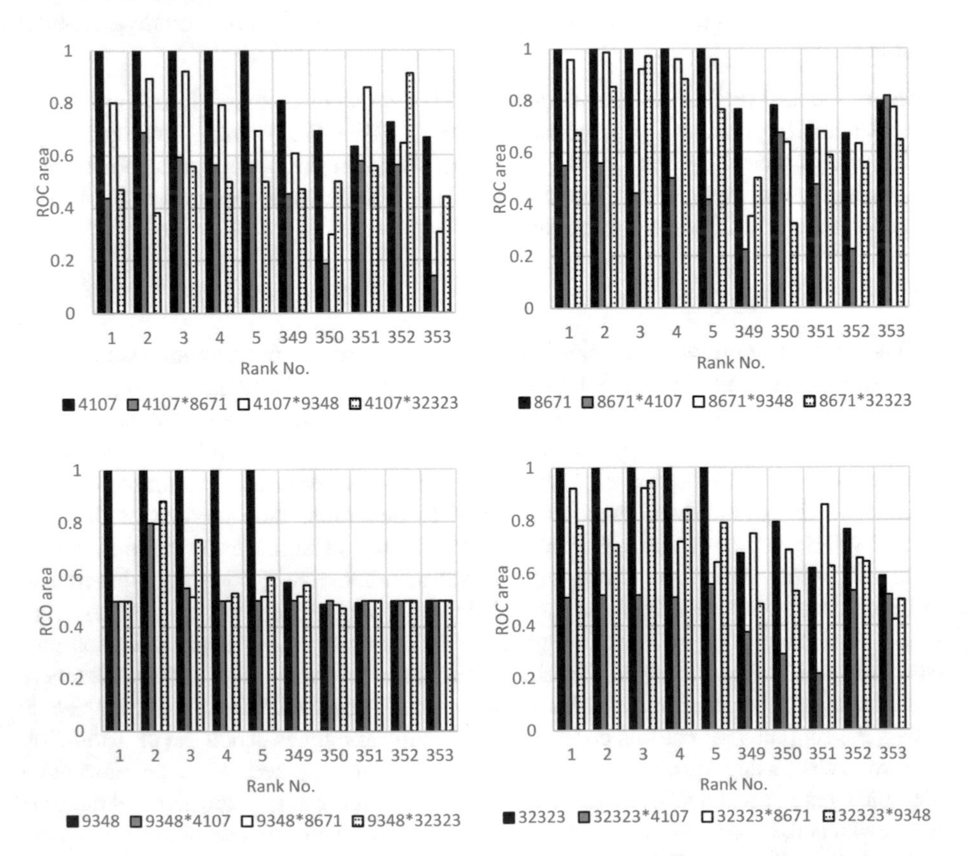

Fig. 1. ROC area of colorectal cancer between top and bottom 5 ranked gene-sets using correlation attribute evaluator.

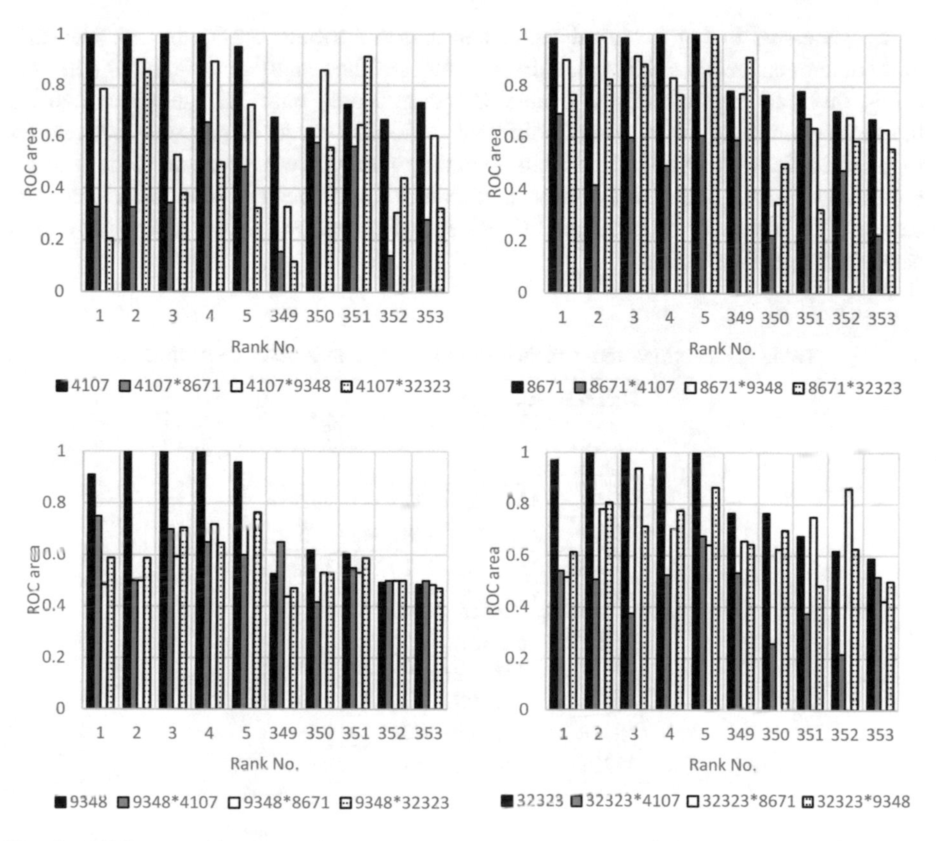

Fig. 2. ROC area of colorectal cancer between top and bottom 5 ranked gene-sets using information gain attribute evaluator.

Table 2. ROC area from using original GSNFS method.

Datasets	ROC area	Datasets	ROC area
4107	1	9348	1
4107_8671	0.547	9348_4107	0.5
4107_9348	0.943	9348_8671	0.5
4107_32323	0.588	9348_32323	0.5
8671	1	32323	1
8671_4107	0.5	32323_4107	0.55
8671_9348	0.951	32323_8671	0.656
8671_32323	0.794	32323_9348	0.833

Next, the consistency of identified top-ranked gene-set based on the occurrence of the same gene-set/pathway across 4 datasets was evaluated. The ranking of gene-sets from using different attribute evaluator can be in a totally different order. The identified top 5 gene-set IDs from the two different attribute evaluators are shown in Table 3.

Many gene-sets in top 5 ranked using correlation attribute evaluator was identified more than once among 4 datasets. However, most of the identified top-ranked gene-sets using information gain attribute evaluator occurred only once. Six gene-sets that has the most occurrence from using 2 different attribute evaluators across 4 datasets are reported in Table 4. TGF-beta receptor signaling pathway was reported to have tumor promoting effects, increasing tumor invasiveness and metastasis in later stages of cancer [17]. Moreover, MAPK and EGFR1 signaling pathways were also known to be cancer related pathways [5].

Table 3. Identified top 5 ranked gene-set ID from 2 attribute evaluators.

Datasets	Rank no.				
	1	2	3	4	5
Correlation					
4107	320	128	221	341	322
8671	282	320	128	341	136
9348	320	341	207	136	128
32323	319	352	207	320	304
InfoGain					
4107	300	8	86	129	249
8671	109	169	61	22	130
9348	109	148	42	40	268
32323	350	282	281	160	361

Table 4. Identified gene-set/pathway and its occurence from using 2 attribute evaluator across 4 datasets.

ID	Occurrence	Pathway
320	4	TGF-beta Receptor Signaling Pathway
128	3	MAPK signaling pathway
341	3	EGFR1 Signaling Pathway
282	2	TNF-alpha/NF-kB Signaling Pathway
136	2	Ubiquitin mediated proteolysis
207	2	Actin Cytoskeleton Signaling

The networks topology of pathway/gene-set with the most occurrences (TGF-beta Receptor Signaling Pathway) from 4 colorectal cancer datasets were analyzed and visualized in Fig. 3. The identified significant genes in each dataset were different and thus resulted in different network topology. The network analysis was reported in Table 5. The network heterogeneity reflects the tendency of a network to contain hub nodes. The network heterogeneity from each dataset after GSNFS is relatively high when compare to the original pathway, which suggest that significant genes were highly connected in the network. The degree of each node was represented by the size

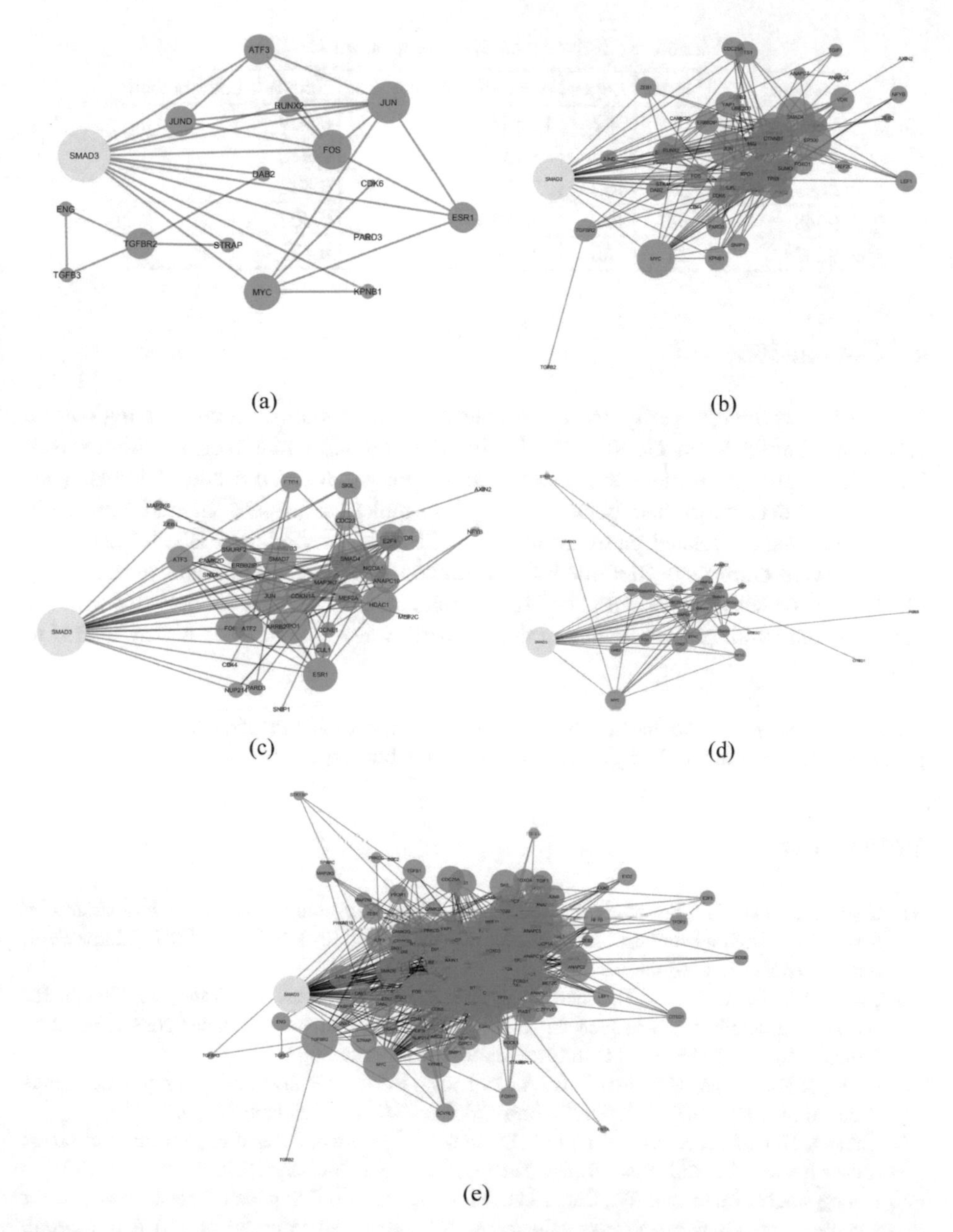

Fig. 3. Gene-set ID 320 network topology obtained from dataset (a) GSE4107, (b) GSE8671, (c) GSE9348, (d) GSE32323, and (e) full original pathway.

of the nodes, whereas bigger node symbolized higher degree of that node. The degree of gene SMAD3 (yellow node) was mostly found to be the highest in the network from 4 different datasets. SMAD3 was reported that it could function as both a tumor suppressor and prometastatic factor [18].

Table 5. Network analysis of gene-set ID 320.

Datasets	Nodes	Edges	Avg. No. of neighbor	Network heterogeneity
Full pathway	147	947	12.884	0.909
4107	16	30	3.75	0.670
8671	50	182	7.28	0.832
9348	39	108	5.538	0.807
32323	28	66	4.714	0.773

4 Conclusion

This work was done to verify and extend our previous studies on incorporating feature selection techniques on GSNFS to identify the suitable candidates of sub-network biomarkers. Similar results were achieved from using colorectal cancer datasets. Top-ranked gene-sets performed better than bottom-ranked gene-sets. In addition, some well-known cancer related pathway such as TGF-beta, MAPK, and EGFR1 signaling pathway were commonly identified in top-ranked gene-sets. Significant genes in top identified gene-set were shown to be highly connected together in the network whereas SMAD3, gene with highest degree in the network, was reported to be highly related to cancer.

Acknowledgements. The first author would like to acknowledge the graduate scholarship from the Department of Chemical Engineering, KMUTT for funding of his Master study.

References

1. Chen, L., Xuan, J., Riggins, R.B., Clarke, R., Wang, Y.: Identifying cancer biomarkers by network-constrained support vector machines. BMC Syst. Biol. **5**(1), 161 (2011). https://doi.org/10.1186/1752-0509-5-161
2. Tyson, J.J., Baumann, W.T., Chen, C., Verdugo, A., Tavassoly, I., Wang, Y., Clarke, R.: Dynamic modelling of estrogen signaling and cell fate in breast cancer cells. Nat. Rev. Cancer **11**(7), 523–532 (2011). https://doi.org/10.1038/nrc3081
3. Curtis, R.K., Orešič, M., Vidal-Puig, A.: Pathways to the analysis of microarray data. Trends Biotechnol. **23**(8), 429–435 (2005). https://doi.org/10.1016/j.tibtech.2005.05.011
4. Chuang, H.Y., Lee, E., Liu, Y.T., Lee, D., Ideker, T.: Network-based classification of breast cancer metastasis. Mol. Syst. Biol. **3**, 140 (2007). https://doi.org/10.1038/msb4100180
5. Doungpan, N., Engchuan, W., Chan, J.H., Meechai, A.: GSNFS: gene subnetwork biomarker identification of lung cancer expression data. BMC Med. Genomics **9**(S3) (2016). https://doi.org/10.1186/s12920-016-0231-4
6. Chan, J.H., Sootanan, P., Larpeampaisarl, P.: Feature selection of pathway markers for microarray-based disease classification using negatively correlated feature sets. In: The 2011 International Joint Conference on Neural Networks, pp. 3293–3299 (2011). https://doi.org/10.1109/ijcnn.2011.6033658
7. Kozuevanich S., Meechai A., Chan J.H.: Feature selection in GSNFS-based marker identification. In: The 10th International Conference on Computational Systems-Biology and Bioinformatics (CSBio 2019). (2019). https://doi.org/10.1145/3365953.3365964

8. Barrett, T.: NCBI GEO: mining millions of expression profiles–database and tools. Nucleic Acids Res. **33**(Database issue), D562–D566 (2004). https://doi.org/10.1093/nar/gki022
9. Soh, D., Dong, D., Guo, Y., Wong, L.: Consistency, comprehensiveness, and compatibility of pathway databases. BMC Bioinform. **11**(1), 449 (2010). https://doi.org/10.1186/1471-2105-11-449
10. Hong, Y., Ho, K.S., Eu, K.W., Cheah, P.Y.: A susceptibility gene set for early onset colorectal cancer that integrates diverse signaling pathways: implication for tumorigenesis. Clin. Cancer Res. **13**(4), 1107–1114 (2007). https://doi.org/10.1158/1078-0432.ccr-06-1633
11. Sabates-Bellver, J., Van der Flier, L.G., de Palo, M., Cattaneo, E., Maake, C., Rehrauer, H., et al.: transcriptome profile of human colorectal adenomas. Mol. Cancer Res. **5**(12), 1263–1275 (2007). https://doi.org/10.1158/1541-7786.mcr-07-0267
12. Hong, Y., Downey, T., Eu, K.W., Koh, P.K., Cheah, P.Y.: A "metastasis-prone" signature for early-stage mismatch-repair proficient sporadic colorectal cancer patients and its implications for possible therapeutics. Clin. Exp. Metas. **27**(2), 83–90 (2010). https://doi.org/10.1007/s10585-010-9305-4
13. Khamas, A., Ishikawa, T., Shimokawa, K., Mogushi, K., et al.: Screening for epigenetically masked genes in colorectal cancer using 5-Aza-2'-deoxycytidine, microarray and gene expression profile. Cancer Genomics Proteomics **9**(2), 67–75 (2012). PMID: 22399497
14. Stark, C.: BioGRID: a general repository for interaction datasets. Nucleic Acids Res. **34** (90001), D535–D539 (2006). https://doi.org/10.1093/nar/gkj109
15. Hall, M.A., Frank, E., Holmes, G., Pfahringer, B., Reutemann, P., Witten, I.H.: The WEKA data mining software: an update. SIGKDD Explor. **11**, 10–18 (2009)
16. Shannon, P.: Cytoscape: a software environment for integrated models of biomolecular interaction networks. Genome Res. **13**(11), 2498–2504 (2003). https://doi.org/10.1101/gr.1239303
17. Syed, V.: TGF-β signaling in cancer. J. Cell. Biochem. **117**(6), 1279–1287 (2016). https://doi.org/10.1002/jcb.25496
18. Millet, C., Zhang, Y.E.: Roles of Smad3 in TGF- β signaling during carcinogenesis. Crit. Rev. Eukaryot. Gene Expr. **17**(4), 281–293 (2009). PMID: 17725494

Smartphone Information Extraction and Integration from Web

Supranee Khamsom(✉) and Wachirawut Thamviset

Department of Computer Science, Faculty of Science, Khon Kaen University, Khonkaen 40002, Thailand
Supranee_K@kkumail.com, twachi@kku.ac.th

Abstract. We present herein a solution to problems in data integration, which is a process of consolidating similar information from different sources, in which multiple data sources ensure data unification. One concept value may have different name values used in two different databases that are consistent and meaningful under the same concept. This conflict must be resolved for consistency as well as to reduce data errors. We extracted the specifications of a mobile phone and smartphone from several websites and created JSON middleware for mapping and synonyms for the specification of mobile phone data in the form of same word standardization. Schema matching plays an important role in combining different sources of information, which can find meaningful consistency between the components of the two schemas, and are then integrated into a new database that collects more mobile phones and smartphones, but reduces the duplication of data from the original database obtained from website data extraction. The application of the proposed method involves the mobile phone data integration problem of two integrated languages, namely, Thai and English, demonstrating efficiency in actual use.

Keywords: Data integration · Middleware · Schema matching

1 Introduction

Currently, the disclosure of various forms of information on the internet becomes problematic for users often in need of information from multiple data sources (websites) [1] that have different names, but similar concepts. This results in both confusion and delays for the user. Each website may have different information and usage of words with different name values, yet with the same meaning under a similar concept [2]; for example, on Website A of the mobile phone, the attribute name of the mobile phone specification is "ข้อมูลเครือข่าย" (network information), whereas the mobile phone on website B names it "เทคโนโลยีเครือข่าย" (network technology). Additionally, on the website of mobile phone A, the attribute name of the mobile phone specification is "แบตเตอรี่มาตรฐาน" ("standard battery"), while the mobile phone containing website B names it "battery", as shown in Fig. 1. In the integration of data from mobile phone specifications, many smartphones may have different names or structures within each website. Each data source, or website, has a different name or structure.

P. Meesad and S. Sodsee (Eds.): IC^{2}IT 2020, AISC 1149, pp. 128–136, 2020.
https://doi.org/10.1007/978-3-030-44044-2_13

Smartphone information from website A	
Name	ออปโป้ OPPO A3s
รุ่นนี้มีขายที่	เปิดตัวครั้งแรก 23 กรกฎาคม 2018 ไตรมาสที่ 3 ปี 2018 ราคามือถือ OPPO A3s ราคาเปิดตัว 4999 บาท กรกฎาคม 61 ราคาล่าสุด 4699 บาท
ข้อมูลตัวเครื่อง	จอแสดงผล IPSLCD 24bit 16 ล้านสี หน้าจอไร้ขอบ Full view Display กว้าง 62 นิ้ว แนวทะแยง ความละเอียด 1520 x 720 พิกเซล Capacitive...
OS CPU	Color OS 51 based on Android 81 Oreo Qualcomm Snapdragon 450 Octa Core ความเร็ว 18 GHz หน่วยความจำ 16 GB ตัวเครื่อง RAM 2GB
ข้อมูลเครือข่าย Network	เครือข่าย GSM 85090018001900 MHz WCDMA 8509002100 MHz เทคโนโลยีการรับส่งข้อมูล 2G EDGEGPRS 3G HSPA 4G NanoSIM
Battery	แบตเตอรี่มาตรฐาน 4230 mAh Standard Battery

Smartphone information from website B	
Name	OPPO A3s
การวางจำหน่าย	วันเปิดตัว Jul 2561 สถานะ วางจำหน่ายแล้ว
หน้าจอแสดงผล	ประเภท IPS LCD ขนาดหน้าจอ 6.20 ความละเอียด 152 x 720 พิกเซล
ประสิทธิภาพ	ระบบปฏิบัติการ Android 8.1 ชิปประมวลผล Qualcomm Snapdragon 450 450 1.8 GHz ชิปกราฟิก Adreno 506 Ram 2 GB ความจุ 16 GB การ์ดหน่วยความจำ รองรับ microSD ความจุที่รองรับ N/A
เทคโนโลยีเครือข่าย	Techn[illegible] HSDPA HSUPA LTE 2G bands 850 / 900 / 1800 / 1900 3G bands 850 / 900 / 2100 4G bands LTE ความเร็ว EDGEGPRS HSDPA HSUPA LTE GPRS Yes EDGE Yes
แบตเตอรี่มาตรฐาน	ประเภทแบตเตอรี่ Li-Polymer ความจุ 4230 mAh

Fig. 1. Examples of data from two websites with synonymous names.

In this manuscript, we propose methods for data extraction from mobile phones employing the DOM tree method, represented through plain text containing HTML tags [3]. Specific keywords are defined in the markup language, which the browser can interpret to display the specific elements of the webpage, as free HTML tags overlap into a hierarchical structure model. In order to modify the structure of the data format from many data sources consistent with the JSON format, JSON has the flexibility to display different data and create middleware for mapping schema and specification data of mobile phones, in which synonyms are presented in a standardized format under the same word. Schema matching plays an important role in combining different sources of information, which can prove meaningful and consistent between the components of the two schemas. Lastly, the databases are integrated in order to reduce the duplication of data from the original database obtained from the initial website's data extraction [4].

2 Related Work

Ahmad et al. [5] presented a method for resolving naming conflicts for the data integration process within XML data sources. This is achieved through the consideration of the attribute name in the data outline of the data dictionary to classify the naming conflicts before matching the attributes of the data schema, in which there are both homonym and synonym conflicts.

Li et al. [6] presented a method of database schema matching using neural network principles, divided into five steps: information enumeration; classification; neural network training; similarity determination; and mapping presentation. This research focused on processes for automatic matching of attributes with the work steps divided into sub-operations, resulting in the matching of the database schema from the source and destination databases, as well as creating the tools and concepts necessary to match the outline.

Su et al. [1] proposed a new solution using XML (eXtensible Markup Language) and middleware technology in which to effectively solve different types of data integration and analysis problems, such as the incompatibility of different data sources. With the development of information technology and the popularity of such networks, even a single department may become a different source of information. The design and creation of a prototype for the general integration framework within a concrete example, via the UIIS (University Integrated Information System), provides a solution to the problem of diversity, as well as the incompatibilities incurred from different data sources.

Chen et al. [7] offered an HDSM schema matching algorithm to combine different data sources that can find meaningful consistency between components of two schemas, complex matching, as well as search for mapping between columns.

3 Methodology

The overall architecture of the proposed system provided the flexibility needed to integrate data from heterogeneous [1] data sources under the same domain, as shown in Fig. 2. Three layers were created: the web extraction layer, integration layer, and the application layer.

3.1 Extraction Web Extraction Layer

In this section, we describe in detail the characteristics of the existing semi-structured web data extraction system [3], in which HTML pages were considered as a form of semi-structured data. In data conversion, data from multiple sources implies the use of different envelopes. Each data source has its own wrapper. The wrapper layer allows the user to receive different structures of the extracted data [1]. The features used to extract web data are the same semi-structural features of the webpage. Within mobile phones, these can be displayed naturally in the form of a tree ordered by the root, where the labels represent the appropriate tags of the syntax, HTML markup language, and tree hierarchy; showing different levels of nesting within the web page elements [3]. To extract the brand name, model, and specification for each mobile device specification; the wrapper will create an XML document to convert the query results to XML [1] using the document object model (DOM), in which page rendering uses a tree root with commands. Represented by plain text with HTML tags, specific keywords are defined in the markup language which the browser interprets in order to display a specific element of a webpage for free text. HTML tags may be nested in a hierarchical

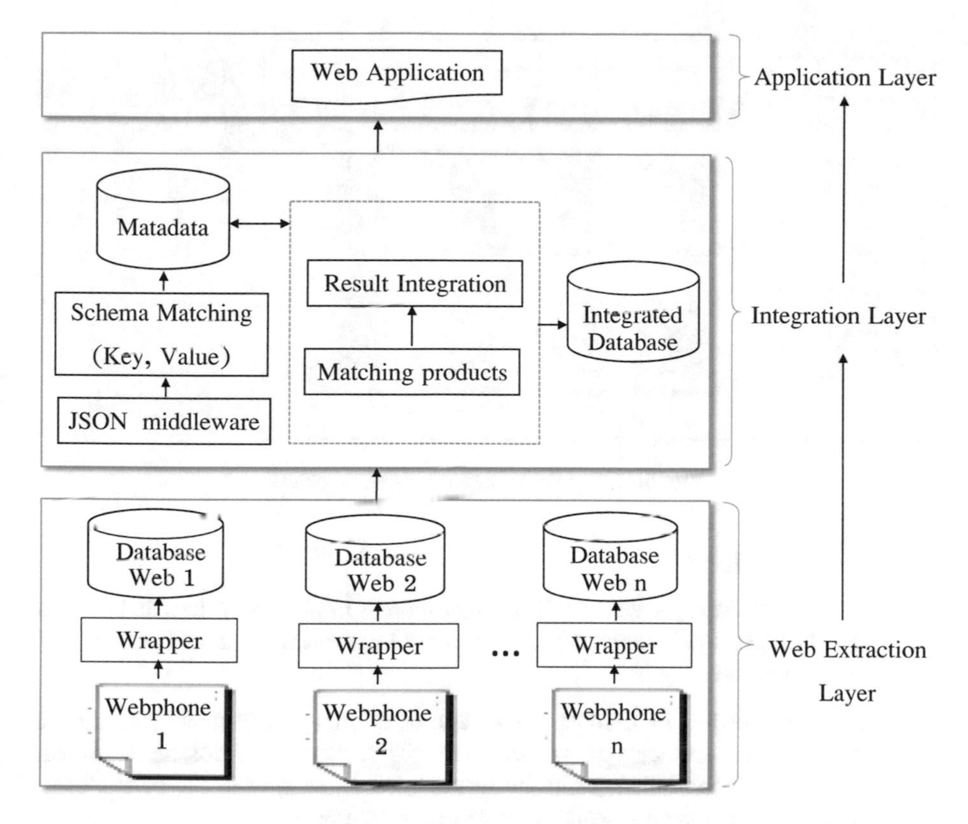

Fig. 2. The architecture of the proposed system.

structure style, which is captured in the DOM by the tree of the document that contains the nodes that represent the HTML tags. The document tree, referred to as the DOM tree, has been frequently utilized for the purpose of web data extraction [3], as shown in Fig. 3.

We addressed the elements in the document tree via XPath. One of the main advantages of using the DOM for web content extraction is the possibility of using certain tools in both the XML and HTML languages. The XML path language provides a powerful syntax for specifying specific elements of an XML document; and, to the same extent, stores the title information of HTML webpages in a simple manner [3]. The brand and model information within the requested specification of each mobile phone model is presented in the form of a key with a value. In order to make the results of data extraction more flexible, the specification of the mobile phone is added later. Ultimately, JSON-formatted search results are sent to the results integrator in the extraction layer to include all search results from the various data sources [1].

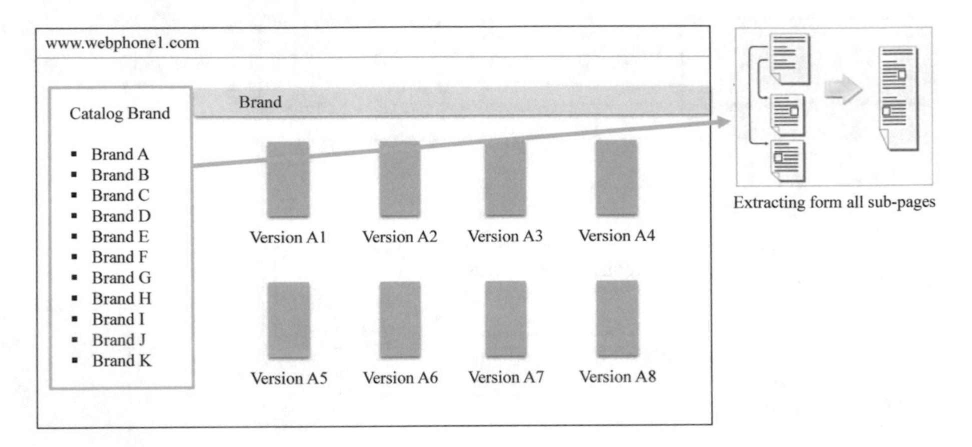

Fig. 3. Examples of mobile phone websites.

3.2 Integration Layer

There is a real difference between multiple data sources in the data values that describe the same [5] object. Such conflicts are considered inconsistent data and contain conflicts, such as with synonyms.

Metadata. Responsible for keeping knowledge of all kinds of different data integration systems. We use metadata to explain the details of the information on the website and to be able to support changes in the form structure, as well as the details of the specifications of smartphones, which change every quarter.

a. JSON middleware. Mobile phone specification data from different sources create value name conflicts. Similarly, meaningful information exists in local databases that have different values; or data that have the same values, yet different meanings. In conflicts with synonyms, data in data cells have the same concept [2], but the names of the data values are different. An important aspect of database integration is having the same concept. However, the names of the data values may differ, resulting in data duplication of the same multiple data sources, requiring the identification and resolution of conflicts through JSON middleware. In an effort to reduce data redundancy, mobile phone specifications of the data source must be to be changed to the same name. This system removes JSON transformations, and merges schemas and local schemas, as shown in Fig. 4.

```
   HeaderStructure={"data":[
{  "key": 1,"value":"Network", "parent": 0},
{  "key": 2,"value":"ข้อมูลเครือข่าย Network", "parent": 1},
{  "key": 3,"value":"เครือข่าย", "parent": 1},

{  "key": 4,"value":"Battery","parent": 0},
{  "key": 5,"value":"แบตเตอรี่มาตรฐาน","parent": 4},
{  "key": 6,"value":"แบตเตอรี่", "parent": 4},

{  "key": 7,"value":"InfoPhone", "parent": 0},
{  "key": 8,"value":"ข้อมูลตัวเครื่อง", "parent": 7},
{  "key": 9,"value":"ตัวเครื่อง", "parent": 7},

{  "key": 10,"value":"Multimedia", "parent": 0},
{  "key": 11,"value":"ฟังก์ชั่นมัลติมีเดีย",  "parent"; 10},
{  "key": 12,"value":"กล้อง", "parent":10},
```


Fig. 4. Examples of middleware.

b. Schema Matching and Mapping information. In the conflict posed by synonyms, we have two different database terms with different names, yet the same values. These values have similar concepts [2] when mapping JSON middleware with the database structure of each website, extracted via the web extraction layer method. After this, the system will adjust the structure of the word conflict, the synonym, from the local database to determine the form of the word structure [6]. In pre-processing [10], we prepare data for the next step of database integration through standardization.

c. Assumptions of word similarity-based mapping information. It's are responsible for predicting a variety of data sources [1] where mobile data specification conflicts occur within multiple data sources from different websites by creating relationships obtained through the mapping process. Within the varied data sources and centralized databases, the corresponding concept will be shown only once in the global schema and is mapped between the specified view and the global schema [7, 11]. Matching is an important step in combining XML schema data obtained from separate data sources, matching operations, or mapping between attributes or values. Restructuring is a final yet optional step in which the global schemas may be analyzed and restructured to remove any redundancies or unnecessary complexity.

Match the Product Name. The data conversion of mobile phone specification allows data from different data sources to be of the same standard. In the next step, we match the products of the same name. We match products of the same model and brand from multiple sources by matching two databases at a time using the first database method (website A obtained from data extraction), which matches the production name of the second database (of website B obtained from data extraction) [8]. Finding the product name corresponding to the mobile phone brand and model will provide the product detail information, or mobile phone specification data, and facilitate the integration into the next step. The production of products with different names indicates that the mobile phones are of different brands or different versions [10]. We collected detailed information on the mobile phone specification products in the integrated database.

Integrated Database. Product data from multiple databases were structured under the same standards as the above process and integrated and collected all smartphone specification products into the integrated database [9] as shown in Table 1. Product information of the same name indicates that the mobile phone is of the same brand and model. After the name is mapped, the system will display the product detail information on the specification of the smartphone and integrating smartphone specification data in the same database as the integrated database. For product information that doesn't find a product with the same name, the system will display the product specification, namely, the smartphone specification, and store the data in the integrated database.

Table 1. Structure of the integrated smartphone database.

Fields	Description
Name	Name, Brand, Model of the mobile phone
Battery	Battery capacity, Type of battery, Charging Standby Time
Connection	Assisted GPS WiFi, Portable WiFi Hotspot, Bluetooth, Type USB
InfoPhone	Display screen details, Display size, Mobile phone size, Colors, Sensor
InfoSell	On sale status, Launch date, Price
Internet	Social network, Application
Messaging	Supported chat messages
Multimedia	Front Camera, Digital camera, Stereo speakers
Network	Data transmission technology
OS	Operating system, CPU, Memory Card
Functions	Organizer, Calendar, Calculator, Alarm clock

3.3 Application Layer

The Application Layer is responsible for displaying user interfaces [1], reporting data from a central database, and displaying data obtained through integration. In order to support the decision-making process, data in the data warehouse must be well organized to meet the needs of different end-users. Users can choose to search for information on demand using a web browser showing the results of the combined data. Note that the user will not know the number of different data sources within the platform. The integration results provide clear and easy to read results on the webpage (Html document).

4 Experimental and Result

Data used in integrating smartphone specification data used a mobile website for extracting the brand, model, mobile phone specification, and all smartphone specification data of each of the three websites; Website A with 576 models, Website B with 312 models, and Website C with 91 models. Therefore, the three mobile phone databases yielded data from multiple sources with different name values under the same meaning, which are synonyms.

In evaluating the effectiveness of the mobile specification schema with JSON middleware for integrating mobile data with the proposed method, we found the number of products with the name, brand, and model of the mobile phone in two databases on Website A which are identical to the database of Website B, Model 149, obtained through system testing and validation. We found the number of products with the name, brand, and model of the mobile phone in two databases on Website A which are identical to the database of Website B, Model 164, by human inspection. Products with different names (brands) were found in 590 different phone models, as shown in Fig. 5. In the experiment, precision is used to evaluate the efficiency and accuracy (91.5%) where the middleware added more synonyms (intelligence) to the system. They may be mapped further to words with different name values under the same meaning for future database integration.

Name	Battery	Battery	Network	Network	Info_Phone	Info_Phone
OPPO F7	แบตเตอรี่มาตรฐาน 3400 mAh Standard Battery	ประเภทแบตเตอรี่ Non-removable Li-Ion ความจุ 3400 m...	เครือข่าย GSM 85090018001900 MHz WCDMA 581 LTE Ban...	Technology EDGE HSPA HSPA+LTE 2G bands 850 / 900 /...	สมาร์ทโฟน โทรศัพท์มือถือพร้อมระบบปฏิบัติการ จอแสดง...	ขนาด 156.00 x 75.30 x 7.80 มม. น้ำหนัก 158 กรัม วั...
OPPO Find X	เทคโนโลยีเพิ่มความเร็วในการชาร์จ แบตเตอรี่มาตรฐาน ...	ประเภทแบตเตอรี่ Non-removable Li-Ion ความจุ 3730 m...	เทคโนโลยีการรับส่งข้อมูล 2G EDGEGPRS 3G HSPA 4G ใช้...	Technology EDGE HSPA HSPA+LTE 2G bands 850 / 900 /...	สมาร์ทโฟน โทรศัพท์มือถือพร้อมระบบปฏิบัติการ จอแสดง...	ขนาด 156.70 x 74.20 x 9.40 มม. น้ำหนัก 186 กรัม วั...
OPPO R15	เทคโนโลยีเพิ่มความเร็วในการชาร์จ แบตเตอรี่มาตรฐาน ...	ประเภทแบตเตอรี่ Non-removable Li-Po Batt ความจุ 34...	เครือข่าย GSM 85090018001900 MHz WCDMA 58421 LTE B...	Technology EDGEHSDPA HSUPALTE 2G bands 850 / 900 /...	สมาร์ทโฟน โทรศัพท์มือถือพร้อมระบบปฏิบัติการ จอแสดง...	ขนาด 155.10 x 75.20 x 7.40 มม. น้ำหนัก 175 กรัม วั...
OPPO R15 PRO	เทคโนโลยีเพิ่มความเร็วในการชาร์จ แบตเตอรี่มาตรฐาน ...	ประเภทแบตเตอรี่ Non-removable Li-Po Batt ความจุ 34...	เครือข่าย GSM 85090018001900 MHz WCDMA 58421 LTE B...	Technology EDGEHSDPA HSUPALTE 2G bands 850 / 900 /...	สมาร์ทโฟน โทรศัพท์มือถือพร้อมระบบปฏิบัติการ จอแสดง...	ขนาด 155.30 x 75.00 x 7.50 มม. น้ำหนัก 175 กรัม วั...
OPPO R17 PRO	เทคโนโลยีเพิ่มความเร็วในการชาร์จ แบตเตอรี่มาตรฐาน ...	ประเภทแบตเตอรี่ Non-removable Li-Po Batt ความจุ 37...	เครือข่าย GSM 85090018001900 MHz WCDMA 658421 LTE ...	Technology HSDPA HSUPALTE 2G bands 850 / 900 / 180...	สมาร์ทโฟน โทรศัพท์มือถือพร้อมระบบปฏิบัติการ จอแสดง...	ขนาด 157.60 x 74.60 x 7.90 มม. น้ำหนัก 183 กรัม วั...
OPPO R9s Pro	เทคโนโลยีเพิ่มความเร็วในการชาร์จ แบตเตอรี่มาตรฐาน ...	ประเภทแบตเตอรี่ Non-removable Li-Po Batt ความจุ 40...	เครือข่าย GSM 85090018001900 MHz WCDMA 85090017001...	Technology EDGEHSDPA HSUPALTE 2G bands 3G bands 85...	สมาร์ทโฟน โทรศัพท์มือถือพร้อมระบบปฏิบัติการ จอแสดง...	ขนาด 163.63 x 80.80 x 7.35 มม. น้ำหนัก 185 กรัม วั...
Samsung Galaxy A8 Star	เทคโนโลยีเพิ่มความเร็วในการชาร์จ แบตเตอรี่มาตรฐาน ...	ประเภทแบตเตอรี่ Non-removable Li-Ion ความจุ 3700 m...	เครือข่าย GSM 85090018001900 MHz UMTS 850900190021...	Technology EDGE HSPA+ LTE Cat.9 VoLTE 3CA 2G bands...	สมาร์ทโฟน โทรศัพท์มือถือพร้อมระบบปฏิบัติการ จอแสดง...	ขนาด 162.40 x 77.00 x 7.60 มม. น้ำหนัก 191 กรัม วั...

Fig. 5. Results of the mobile integrated web.

5 Conclusion

To solve the problems associated with heterogeneous data source integration of smartphone specification data, we introduced the JSON middleware for data mapping. As a central part of the data mapping, the smartphone specifications of all data sources are converted into the standard names. This method reduces manual work on data-name conversion and data consolidation, is very efficient, and provides users with a logical view of the data. For synonym conflict resolution, the databases extracted from multiple websites will be aligned, that the same concept data cells can be merged into a single data cell thereby reducing data redundancy. A possible direction of future research is to conduct recognition and normalization jointly, such that the integrated data from each extracted website helps to build a wrapper and JSON middleware for extracting and integrating from other websites.

Acknowledgment. This research was partially supported by the Department of Computer Science, Faculty of Science, Khon Kaen University, Khon Kaen, Thailand.

References

1. Su, J., Fan, R., Li, X.: Research and design of heterogeneous data integration middleware based on XML. In: 2010 IEEE International Conference on Intelligent Computing and Intelligent Systems, pp. 850–854 (2010)
2. Mirza, G.A.: Value name conflict while integrating data in database integration. In: 2014 11th International Computer Conference on Wavelet Active Media Technology and Information Processing (ICCWAMTIP), pp. 316–320 (2014)
3. Ferrara, E., De Meo, P., Fiumara, G., Baumgartner, R.: Web data extraction, applications and techniques: a survey. Knowl.-Based Syst. **70**, 301–323 (2014). https://doi.org/10.1016/j.knosys.2014.07.007
4. Sangkla, K., Seresangtakul, P.: Information integration of heterogeneous medical database systems using metadata. In: 2017 21st International Computer Science and Engineering Conference (ICSEC), pp. 1–5 (2017)
5. Ahamed, B.B., Ramkumar, T., Hariharan, S.: Data integration progression in large data source using mapping affinity. In: 2014 7th International Conference on Advanced Software Engineering and Its Applications, pp. 16–21 (2014)
6. Li, Y., Liu, D.-B., Zhang, W.-M.: Schema matching using neural network. In: The 2005 IEEE/WIC/ACM International Conference on Web Intelligence (WI 2005), pp. 743–746 (2005)
7. Chen, W., Guo, H., Zhang, F., Pu, X., Liu, X.: Mining schema matching between heterogeneous databases. In: 2012 2nd International Conference on Consumer Electronics, Communications and Networks (CECNet), pp. 1128–1131 (2012)
8. Ahmad, K., Chiew, H.K., Samad, R.: Intelligent Schema Integrator (ISI): a tool to solve the problem of naming conflict for schema integration. In: Proceedings of the 2011 International Conference on Electrical Engineering and Informatics, pp. 1–5 (2011)
9. Gou, H., Jing, Y., Feng, B., Li, Y.: A scheme of information integration based on XML description and schema matching. In: 2012 Fourth International Conference on Computational and Information Sciences, pp. 381–384 (2012)
10. Melnik, S., Garcia-Molina, H., Rahm, E.: Similarity flooding: a versatile graph matching algorithm and its application to schema matching. In: Proceedings 18th International Conference on Data Engineering, pp. 117–128 (2002)
11. Madhavan, J., Bernstein, P., Chen, K., Halevy, A., Shenoy, P.: Corpus-based schema matching. In: In ICDE, pp. 57–68 (2003)

A Comparative Study on Artificial Neural Network and Radial Basis Function for Modelling Output Response from Computer Simulated Experiments

Anamai Na-udom[1(✉)] and Jaratsri Rungrattanaubol[2]

[1] Department of Mathematics, Faculty of Science, Naresuan University, Phitsanulok, Thailand
anamain@nu.ac.th

[2] Department of Computer Science and Information Technology, Faculty of Science, Naresuan University, Phitsanulok, Thailand
jaratsrir@nu.ac.th

Abstract. Computer simulated experiments (CSE) have been widely used to investigate complex physical phenomena, particularly when physical experiments are not feasible due to limitations of experimental materials. The natures of CSE are time-consuming and the computer codes are expensive. Therefore, experimental designs and statistical models approaches play a major role in the context of CSE in order to develop the approximation model for use as a surrogate model. Many researchers have attempted to develop various predictive models to fit the output responses from CSE. The purpose of this paper is to compare the prediction accuracy of three models namely Kriging model (KRG), Radial basis function (RBF) model and Artificial neural network (ANN) model, respectively. These three models are constructed by using the optimal Latin hypercube designs (OLHD). The prediction accuracy of each model is validated though non-linear test problems ranging from 2 to 8 input variables and evaluated by the root mean squared of error (RMSE) values. The results show that RBF model performs well when small dimension of problem with small design run is considered while KRG model is the most accurate model when the design run is large. For larger dimensions of problem, KRG model is suitable for small design runs while ANN model performs superior over the other models when the design runs are large.

Keywords: Artificial neural network · Predictive model · Radial basis functions · Computer simulated experiments

1 Introduction

Computer simulated experiments (CSE) have been practiced in various fields such as petroleum engineering, mining industrial and applied science to explore the complicated phenomena, especially when physical experiments are not feasible due to time and cost constraints or insufficient of experimental units. The examples of CSE are the use of computational fluid dynamics (CFD) model to study the oil mist separator

P. Meesad and S. Sodsee (Eds.): IC^{2}IT 2020, AISC 1149, pp. 137–148, 2020.
https://doi.org/10.1007/978-3-030-44044-2_14

system in the internal combustion engine, a finite element model for simulating frontal crashes to develop vehicle structure, a study of environmental pollutants that affect human health [1] and so on. The nature of CSE is that the output response is deterministic as the same settings of input variables will always provide the same value of output response. Hence, replication, blocking and randomization that play very important role in physical experiment are not necessary for CSE [2]. Generally, the space filling designs such as Latin hypercube design or uniform design that aim to spread the design points over a region of interests are normally practiced in CSE. Further, running computer codes are time-consuming and expensive so the constructions of predictive model have been developed by many researchers in this context to overcome these problems.

Kriging model has been widely used for modeling output response from CSE due to its interpolation property which is completely accurate when the prediction point is close to the design point [3]. The drawback of Kriging model is that the estimation of all unknown parameters is based on the optimization method and sometimes the optimal settings of parameters are not achieved and hence the prediction accuracy of Kriging model becomes worse. There have been a range of research publications that focus on enhancement of the parameter estimation method for estimating unknown parameters in Kriging model. For instance, Welch et al. [4] proposed an efficient algorithm to estimate the parameters using the maximum likelihood method.

While Kriging model has received wide attention in developing the predictive models, there are some statistical models such as response surface methodology (RSM), Multivariate adaptive regression splines (MARS), Radial basis function (RBF) have been adopted to use in the context of CSE. Various published works have attempted to compare the prediction accuracy of models for predicting the output responses from CSE. For instance, Simpson et al. [5] compared the performance of Kriging models and RSM in aerospace engineering application. The results showed that Kriging models were slightly more accurate than RSM. Jin et al. [6] studied the prediction accuracy of four different models, polynomial regression, Kriging, MARS and RBF using Latin Hypercube designs. The results showed that RBF model was the most accurate model. Hussain et al. [7] compared the prediction accuracy between polynomial models and various forms of basis function of RBF using factorial designs and Latin hypercube designs. The results revealed that RBF models provide higher accuracy than polynomial models for all test problems. Fang and Horstemeyer [8] studied the performance of RBF and RSM and the result showed that RBF models performed better than RSM. Mullur and Messac [9] investigated the prediction accuracy among various types of RBF, RSM and Kriging models using different classes of designs. The results indicated that RBF model was comparable to Kriging model. Yosboonruang et al. [10] compared the prediction accuracy among Kriging models, RSM and RBF using optimal Latin hypercube design (OLHD), and two classical designs namely Central composite design (CCD) and Fractional factorial design (FFD). The results indicated that RSM performed best when the optimal Latin hypercube design was used with non-complex problem. In the case of complex problem, Kriging models and RBF models were superior over RSM. Furthermore, when the classical designs were considered, RBF model performed best. Na-udom and Rungrattanaubol [11] compared the performance of Kriging and ANN models and the results showed that ANN performed well and can be used as an alternative to Kriging model in some features of problem. Vicario et al. [12] compared the performance of Kriging and ANN models for predicting the response using four-dimensional computational fluid dynamics experiments. The results showed that Kriging and ANN are comparable in terms of prediction accuracy.

According to the results published so far, it could be noticed that Kriging, RBF and ANN performs better than other models in various case studies. Further, there are still no certain conclusions on which predictive model is the best choice to use for any specific problems in the context of CSE. Therefore, this paper aims to compare the prediction accuracy between the three modeling methods namely KRG, RBF and ANN models based on small and large design runs. The optimal Latin hypercube designs (OLHD) are generates by optimization algorithm and used to construct the models. The prediction accuracy of each model is validated through non-linear test problems ranging from 2 to 8 input variables and evaluated by the root mean squared of error (RMSE) values. In Sect. 2, we will present the research method which consists of details of the three predictive models and test problems. The results will be delivered in Sect. 3 and the conclusions will be presented in the Sect. 4, respectively.

2 Research Methods

In this section, we present the details of how to construct the predictive models, Kriging, RBF, and ANN, respectively. The details of each test problems are also described and the model validation method such as root mean squared of error and percentage improvement are also given.

2.1 Kriging Model

Kriging model has been originally used in mining engineering and geostatistics [13]. The model has received attention in applications of computer simulated experiments because of its interpolation property. Sacks et al. [3] adopted Kriging model for CSE and the mathematical form of this model can be expressed as,

$$y = \sum\nolimits_{j=1}^{d} \beta_j f_j(\boldsymbol{x}) + Z(\boldsymbol{x}), \tag{1}$$

where d is the number of input variables.

The form of Kriging model is based on the idea that the output response can be modeled as the combination of a polynomial function of input variables and a realization of stochastic process, $Z(\boldsymbol{x})$, with zero mean and a form of correlation function given by

$$Cov\left[Z(\boldsymbol{x}_i), Z(\boldsymbol{x}_j)\right] = \sigma^2 R(\boldsymbol{x}_i, \boldsymbol{x}_j) \tag{2}$$

where σ^2 is the process variance and R is the correlation between two design points x_i and x_j.

Various forms of correlation functions were proposed for Kriging model. The Guassian form is the most frequently used and can be written as,

$$R(\boldsymbol{x}_i, \boldsymbol{x}_j) = \prod_{l=1}^{d} exp\left(-\theta_l \left|x_i^{(l)} - x_j^{(l)}\right|^p\right) \tag{3}$$

where $\theta_l > 0$ and $0 < p \leq 2$. In this study, we set $p = 2$.

The polynomial function part in Eq. (1) can be replaced by a constant vector of 1 as the prediction accuracy of Kriging model would not be significantly affected [3, 4]. Therefore, the subsequent Kriging model can be expressed as,

$$y = \beta + Z(\boldsymbol{x}) \tag{4}$$

All unknown parameters in the correlation function, θ, can be estimated by an algorithm based on the maximum likelihood estimation method (MLE) proposed by Welch et al. [4]. Further, the estimators of parameter, $\boldsymbol{\beta}$, and process variance, σ^2, can be obtained by maximizing the log likelihood function as follows,

$$l(\boldsymbol{\beta}, \sigma^2, \theta) = -\frac{1}{2}\left[n\ln\sigma^2 + \ln|\boldsymbol{R}| + (\boldsymbol{y} - \mathbf{1}\boldsymbol{\beta})^T \boldsymbol{R}^{-1}(\boldsymbol{y} - \mathbf{1}\boldsymbol{\beta})/\sigma^2\right], \tag{5}$$

where $\boldsymbol{y}$ is the column vector of length n that contains the true response at each design point, $\boldsymbol{R}$ is the $n \times n$ symmetric correlation matrix, $R(\boldsymbol{x}_i, \boldsymbol{x}_j)$ for any two design points $(1 \leq i, j \leq n)$.

Given the correlation parameter, θ, into Eq. (3), the generalized least square estimator of $\boldsymbol{\beta}$ is,

$$\widehat{\boldsymbol{\beta}} = \left(\mathbf{1}^T \boldsymbol{R}^{-1} 1\right)^{-1} 1^T \boldsymbol{R}^{-1} \boldsymbol{y} \tag{6}$$

and the MLE of σ^2 is

$$\widehat{\sigma}^2 = \frac{1}{n}\left(\boldsymbol{y} - \mathbf{1}\hat{\boldsymbol{\beta}}\right)^T \boldsymbol{R}^{-1}\left(\boldsymbol{y} - \mathbf{1}\,\widehat{\boldsymbol{\beta}}\right) \tag{7}$$

Substituting $\widehat{\boldsymbol{\beta}}$ and $\widehat{\sigma}^2$ into the likelihood function in Eq. (5), the optimal setting of correlation parameter can be estimated by maximizing the following equation.

$$-\frac{1}{2}\left(n\ln\widehat{\sigma}^2 + \ln|\boldsymbol{R}|\right), \tag{8}$$

which is a function of the correlation function parameters and the collected data.

After all parameters are estimated, the next process is to construct a predictor, $\widehat{y}(\boldsymbol{x})$, of $y(\boldsymbol{x})$ to act as an approximate model for the output response from CSE. The best linear unbiased predictor (BLUP) at an untried input $\boldsymbol{x}$ is

$$\hat{y}(\boldsymbol{x}) = \hat{\boldsymbol{\beta}} + \boldsymbol{r}^T(\boldsymbol{x})\boldsymbol{R}^{-1}\left(\boldsymbol{y} - 1\hat{\boldsymbol{\beta}}\right) \tag{9}$$

where $\boldsymbol{r}^T(\boldsymbol{x}) = [\,R(\boldsymbol{x}_1, \boldsymbol{x}) \quad \ldots \quad R(\boldsymbol{x}_n, \boldsymbol{x})\,]$ is the vector of correlation function between n design points and untried input $\boldsymbol{x}$.

In this paper, all unknown correlation parameters are estimated using the method enhanced by Na-udom and Rungrattanaubol [14].

2.2 Radial Basis Function

Radial basis functions (RBF) were originally developed to fit irregular topographic contours of geographical data. The method is basically based on the interpolation of data in multi-dimension problems [8], especially when the number of the observations or design run is large. RBF uses a linear combination of a radially symmetric function based on Euclidean distance or other distances metric to approximate the response functions [6]. A mathematical form of RBF model can be expressed as,

$$y(\boldsymbol{x}) = \sum_{i=1}^{n} \beta_i \phi(\|\boldsymbol{x} - \boldsymbol{x}_i\|) \tag{10}$$

where $\boldsymbol{n}$ is the number of design runs, $\boldsymbol{x}$ is a vector of input variables, $\boldsymbol{x}_i$ is a vector of input variables at the $\boldsymbol{i}^{th}$ design run, $\|\boldsymbol{x} - \boldsymbol{x}_i\|$ is the Euclidean norm, $\boldsymbol{\phi}$ is a basis function, and $\boldsymbol{\beta}_i$ is the coefficient for the $\boldsymbol{i}^{th}$ basis function. A form of basis function that normally uses is presented in Table 1. In this study, we conduct an empirical study to find the optimal choice of basis function for constructing RBF model.

Table 1. Basis functions for RBF model.

Name	Basis Function
Gaussian	$\phi(\|x - x_i\|) = e^{-c\|x-x_i\|^2}, 0 < c \leq 1$
Multiquadric	$\phi(\|x - x_i\|) = \sqrt{\|x - x_i\|^2 + c^2}, 0 < c \leq 1$
Inverse multiquadric	$\phi(\|x - x_i\|) = \frac{1}{\sqrt{\|x-x_i\|^2 + c^2}}, 0 < c \leq 1$
Thin-plate spline	$\phi(\|x - x_i\|) = \|x - x_i\|^2 \ln(\|x - x_i\|)$
Linear	$\phi(\|x - x_i\|) = \|x - x_i\|$
Cubic	$\phi(\|x - x_i\|) = \|x - x_i\|^3$

Replacing $\boldsymbol{x}$ and $\boldsymbol{y}(\boldsymbol{x})$ in Eq. (10) with the $\boldsymbol{n}$ vectors of input variables and corresponding function values leads to the following equations,

$$\begin{aligned}
y(\boldsymbol{x}_1) &= \sum_{j=1}^{n} \beta_j \phi\left(\|\boldsymbol{x}_1 - \boldsymbol{x}_j\|\right) \\
y(\boldsymbol{x}_2) &= \sum_{j=1}^{n} \beta_j \phi\left(\|\boldsymbol{x}_2 - \boldsymbol{x}_j\|\right) \\
&\vdots \\
y(\boldsymbol{x}_n) &= \sum_{j=1}^{n} \beta_j \phi\left(\|\boldsymbol{x}_n - \boldsymbol{x}_j\|\right)
\end{aligned}$$

The above equations can be written in matrix form as,

$$\boldsymbol{y} = \boldsymbol{F}\boldsymbol{\beta} \tag{11}$$

where $\boldsymbol{F}_{i,j} = \phi(\|\boldsymbol{x}_i - \boldsymbol{x}_j\|)(i,j = 1,2,\ldots,n)$, $\boldsymbol{\beta} = [\beta_1\beta_2\ldots\beta_n]^T$ and $\boldsymbol{y} = [y(\boldsymbol{x}_1)y(\boldsymbol{x}_2)\ldots y(\boldsymbol{x}_n)]^T$

Hence the least squares estimator of $\boldsymbol{\beta}$ can be estimated by using the following equation.

$$\widehat{\boldsymbol{\beta}} = \left(\boldsymbol{F}^T\boldsymbol{F}\right)^{-1}\boldsymbol{F}\boldsymbol{y} \tag{12}$$

It should be noted that if Gaussian, Multiquadric or Inverse multiquadric have been used, the parameter c must be estimated prior to fit RBF model. In this study, we use the method proposed by Rippa [15] as it provides the optimal c that minimizes the interpolation error.

2.3 Artificial Neural Network

Artificial neural network (ANN) is non-parametric approach which any assumption is not required prior to fit the model. The method has been applied successfully in various fields such as data mining, engineering, medicine, finance [16] and so on. The inspiration for neural networks was the recognition that complex learning systems in animal brains consisted of closely interconnected sets of neurons. The structure and process of a neuron is that the dendrites collect the inputs from other neurons to cell body, then combines the input information in order to form a nonlinear relationship between input and output response and then send a functional form to other neurons by axon.

A process of simple artificial neuron model consists of input variable (x_i) multiplied by weight (w_i) associated with the ith input variable, then the obtained products are combined by a summation function $(\sum)$. The summation point is normally referred as a node, and bias (b) is given to each node. The output from a node is passed to a transfer function or an activation function (f) to produce the output response (y) as shown in Fig. 1.

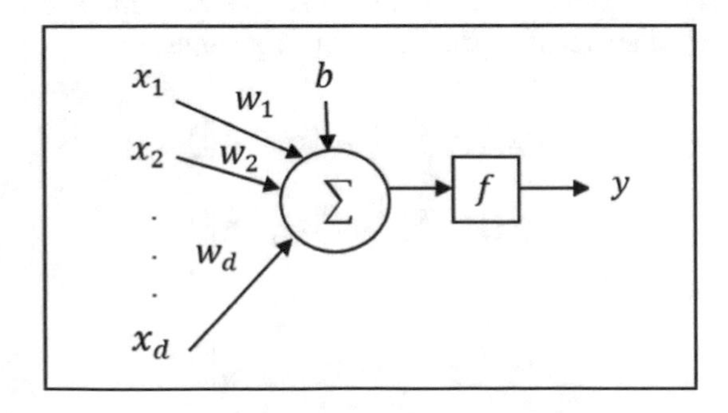

Fig. 1. A simple artificial neuron model.

The process of artificial neuron model can be mathematically written as

$$y = f\left(\sum_{i=1}^{d} x_i w_i + b\right), \tag{13}$$

where d is the number of input variables.

The most frequently used activation function is sigmoid function that can be expressed as,

$$f(a) = \frac{1}{1 + e^{-a}} \tag{14}$$

A structure of ANN has three main layers: input layer, hidden layer and output layer as shown in Fig. 2. The number of hidden layers and the number of nodes in each hidden layer are both configurable by user. ANN is completely connected network, every node in a previous layer is connected to every node in the next layer by associated weight [17].

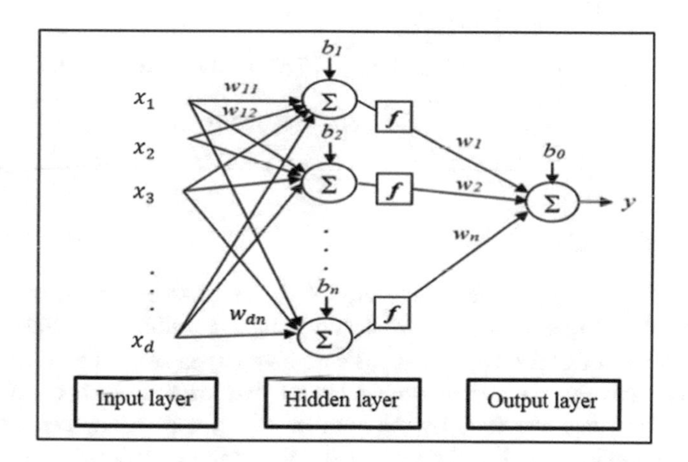

Fig. 2. A structure of ANN.

The most popular learning algorithm given to train the data is back propagation. By using this method, the prediction errors are the difference between the actual values and the predicted values, are fed back through the network. The weights on all connections are adjusted in order to reduce the error by using gradient descent method. The method involves learning rate and momentum rate ranging from 0 to 1. A learning rate affects how large the weight adjustment should be set and the momentum rate influences the adjustment in the current weight to move along the same direction as previous adjustments.

In this study, we set learning rate, momentum rate, training time, the number of hidden layers and their nodes in Weka 3.8 to construct the ANN models.

2.4 Test Problems

The test problems used to validate model are obtained from [3, 4, 18] and the details on range of all input variables and the features of each function are given in Table 2.

Table 2. The detail of test functions.

Test function	d	Function
Branin	2	$y = \left(x_2 - \frac{5.1}{4\pi^2}x_1^2 + \frac{5}{\pi}x_1 - 6\right)^2 + 10\left(1 - \frac{1}{8\pi}\right)\cos(x_1) + 10,$ $-5 \le x_1 \le 10, \quad 0 \le x_2 \le 15$
Welch	2	$y = [30 + x_1 \sin(x_1)](4 + e^{-x_2}), 0 \le x_1, x_2 \le 5$
Cyclone	7	$y = 174.42\left(\frac{x_1}{x_5}\right)\left(\frac{x_3}{x_2 - x_1}\right)^{0.85}\sqrt{\frac{1 - 2.62\left\{1 - 0.36(x_4/x_2)^{-0.56}\right\}^{3/2}(x_4/x_2)^{1.16}}{x_6 x_7}}$ $0.09 \le x_1, x_3, x_4 \le 0.11, 0.27 \le x_2 \le 0.33, 1.35 \le x_5 \le 1.65,$ $14.4 \le x_6 \le 17.6, 0.675 \le x_7 \le 0.825$
Borehole	8	$y = \frac{2\pi x_3(x_4 - x_6)}{\ln\left(\frac{x_2}{x_1}\right)\left[1 + \frac{2x_7 x_3}{\ln\left(\frac{x_2}{x_1}\right)x_1^2 x_8} + \frac{x_3}{x_5}\right]}$ $0.05 \le x_1 \le 0.15, 100 \le x_2 \le 50000, 63070 \le x_3 \le 115600,$ $990 \le x_4 \le 1110,$ $63.1 \le x_5 \le 116, 700 \le x_6 \le 820, 1120 \le x_7 \le 1680,$ $9855 \le x_8 \le 12045$

2.5 Model Validation

In order to validate the prediction accuracy of the predictive models, we generate 10 different OLHD designs using simulated annealing algorithm (SA) under ϕ_p criteria [19]. The number of dimensions or input variables (d) is set to be 2, 7, and 8 input variables, respectively. Each dimension includes two sizes of design runs (n), small and large design runs, classified by the number of terms of parameters required in a quadratic polynomial model, calculated by the following equation.

$$n = 2d + 4\binom{d}{2} + 1 \tag{15}$$

The prediction accuracy of each model is validated using various non-linear test problems. A comparative study is performed by using a root mean squared error (RMSE) values as follows,

$$RMSE = \sqrt{\frac{\sum_{i=1}^{m}(y_i - \hat{y}_i)^2}{m}}, \tag{16}$$

where m is the number of test point, y_i is the actual response of the ith test point and $\hat{y}_i$ is the predicted value of output response from the models at the ith test point.

Moreover, in order to scale the magnitude of RMSE values from various types and ranges of test problems, we calculate the percentage improvement (PI) of ANN and RBF over Kriging model defined as,

$$PI = \frac{RMSE(Kriging) - RMSE(RBF, ANN)}{RMSE(Kriging)} \times 100\% \quad (17)$$

3 Results

In this section, the prediction accuracy of Kriging, the optimal RBF and ANN models are compared by RMSE values. The different OLHD designs are generated for 10 times for each dimension and number of design runs. The minimum, maximum, mean, and standard deviations of RMSE values over 10 replications are also presented. Further, percentage improvement (PI) values over Kriging model are also shown in Tables 3, 4, 5 and 6 respectively.

Table 3. RMSE values for Branin function (d = 2).

n	Model	RMSE				
		Min	Max	Mean	PI (%)	S.D.
6	KRG	41.472	57.585	47.527		5.290
	RBF	41.528	57.825	**45.690**	3.040	5.533
	ANN	58.388	63.878	56.478	−28.329	1.744
9	KRG	21.838	40.808	**31.323**		9.998
	RBF	21.715	41.626	31.671	−1.110	10.494
	ANN	56.849	58.596	47.794	−84.512	0.622

According to all statistics presented in Table 3, it can be concluded that RBF performs best when the design run is small, followed by KRG model. When the design run becomes large, KRG and RBF models are comparable while KRG is slightly better than RBF. It should be noted from S.D. values that KRG is more consistent than RBF model.

It can be clearly seen from Table 4 that RBF performs better than other models when the design run is small and when the design run becomes larger, KRG model turns to be the best choice as the mean of RMSE values obtained from KRG are lower than that of RBF and ANN models. Hence it can be concluded that in the cases of small number of input variables with complex feature of output response, RBF model is recommended for small design run while KRG model seems to be the best choice to use when the design run is large.

Table 4. RMSE values for Welch function (d = 2).

n	Model	RMSE				
		Min	Max	Mean	PI (%)	S.D.
6	KRG	4.483	6.439	5.492		0.808
	RBF	3.479	6.499	**5.170**	5.874	1.339
	ANN	8.404	9.807	9.020	−64.237	0.498
9	KRG	3.382	4.326	**3.854**		0.497
	RBF	4.043	4.376	4.210	−9.224	0.175
	ANN	8.253	8.875	8.493	−120.359	0.194

Cyclone model has been extensively used to validate the performance of predictive model in the context of CSE. Table 5 shows that KRG model performs best when the design run is small while both of KRG and ANN models are superior over RBF. The S. D. values also confirm that KRG and ANN are robust to different designs used to fit the predictive model.

Table 5. RMSE values for Cyclone model (d = 7).

n	Model	RMSE				
		Min	Max	Mean	PI (%)	S.D.
36	KRG	0.004	0.006	**0.005**		0.001
	RBF	0.046	0.077	0.059	−1155.319	0.010
	ANN	0.007	0.010	0.009	−97.021	0.001
99	KRG	0.004	0.005	**0.004**		0.000
	RBF	0.009	0.016	0.012	−174.419	0.002
	ANN	0.004	0.005	**0.004**	0.000	0.000

Borehole function has been widely used to implement the accuracy of the predictive model. All statistics presented in Table 6 reveal that KRG model performs best when the design run is small. PI values over KRG of other models are the large negative quantity. This indicates that KRG is far better than other models. For large design runs, RMSE values from ANN model are lower than other models. Hence ANN is recommended for use to develop a predictive model. It should be noted from PI values that the performance of KRG model is very close to ANN model. Hence ANN is recommended for use to develop a predictive model.

Table 6. RMSE values for Borehole function (d = 8).

n	Model	RMSE				
		Min	Max	Mean	PI (%)	S.D.
45	KRG	0.742	1.079	**0.949**		0.106
	RBF	4.456	5.514	5.067	−433.997	0.342
	ANN	1.243	3.058	2.046	−115.583	0.620
129	KRG	0.647	1.069	0.840		0.140
	RBF	2.456	3.369	2.897	−245.027	0.286
	ANN	0.614	0.864	**0.710**	15.426	0.083

4 Conclusions

This paper aims to compare the prediction accuracy of Kriging, Radial basis function and Artificial Neural Network models for modelling output response from CSE. According to the results presented in the previous section, it indicates that RBF is suitable for small dimensions of problems with small design run while KRG is the most accurate model in a case of large design runs. For large dimensions with complex feature, the results reveal that KRG model is the best choice to construct a predictive model for CSE especially when small design runs are used. In addition, ANN model performs best when the design run becomes large. Hence, ANN model is recommended for constructing the predictive model of output response from CSE, especially when the dimension of problem and design run are large.

References

1. Fang, H., Horstemeyer, M.F.: Global response approximation with radial basis functions. Eng. Optim. **38**(4), 407–424 (2006)
2. Simpson, T.W., Mauery, T.M., Korte, J.J., Mistree, F.: Kriging models for global approximation in simulation-based multidisciplinary design optimization. AIAA J. **39**(12), 2233–2241 (2001)
3. Sacks, J., Welch, W.J., Mitchell, T.J., Wynn, H.P.: Design and analysis of computer experiments. Stat. Sci. **4**(4), 409–435 (1989)
4. Welch, W.J., Buck, R.J., Sacks, J., Wynn, H.P., Mitchell, T.J., Morris, M.D.: Screening, predicting, and computer experiments. Technometrics **34**(1), 15–25 (1992)
5. Simpson, T.W., Lin, D.K.J., Chen, W.: Sampling strategies for computer experiments: design and analysis. Int. J. Reliab. Appl. **2**(3), 209–240 (2001)
6. Jin, R., Chen, W., Simpson, T.W.: Comparative studies of metamodeling techniques under multiple modeling criteria. Struct. Multidiscip. Optim. **23**, 1–13 (2001)
7. Hussian, M.F., Barton, R.R., Joshi, S.B.: Metamodeling: radial basis functions, versus polynomials. Eur. J. Oper. Res. **138**, 142–154 (2002)
8. Fang, K.T., Li, R., Sudjianto, A.: Design and Modeling for Computer Experiments. Chapman & Hall/CRC, London (2006)
9. Muller, A.A., Messac, A.: Metamodeling using extended radial basis functions: a comparative approach. Eng. Comput. **21**, 203–217 (2006)

10. Yosboonruang, N., Na-udom, A., Rungrattanaubol, J.: A comparative study on predicting accuracy of statistical models for modeling deterministic output responses. Thailand Stat. **11** (1), 1–15 (2013)
11. Na-udom, A., Rungrattanaubol, J.: A comparison of artificial neural network and Kriging model for predicting the deterministic output response. NU Sci. J. **10**(1), 1–9 (2014)
12. Vicario, G., Craparotta, G., Pistone, G.: Meta-models in computer experiments: Kriging versus Artificial Neural Networks. Qual. Reliab. Eng. Int. **32**(6), 2055–2065 (2016)
13. Cressie, N.A.C.: Statistics for Spatial Data. Wiley, Hoboken (1993)
14. Na-udom, A., Rungrattanaubol, J.: Optimization of correlation parameter for Kriging approximation model. In: International Joint Conference on Computer Science and Software Engineering, vol. 1, pp. 159–164 (2008)
15. Rippa, S.: An algorithm for selecting a good value for the parameter *c* in radial basis function interpolation. Adv. Comput. Math. **11**, 193–210 (1999)
16. Sibanda, W., Pretorius, P.: Artificial neural networks-a review of applications of neural networks in the modeling of HIV epidemic. Int. J. Comput. Appl. **44**, 1–9 (2012)
17. Larose, D.T., Larose, C.D.: Discovering Knowledge in Data: An Introduction to Data Mining, 2nd edn. Wileys, Hoboken (2014)
18. Hock, W., Schittkowski, K.: Test Examples for Nonlinear Programming Codes. Springer, Berlin (1981)
19. Na-udom, A., Rungrattanaubol, J.: Heuristic search algorithms for constructing optimal latin hypercube designs. In: Recent Advances in Information and Communication Technology, vol. 463, no. 1, pp. 183–193 (2016)

Mutation Variations in Improving Local Optima Problem of PSO

Ekkarat Adsawinnawanawa and Boontee Kruatrachue(✉)

Department of Computer Engineering, Faculty of Engineering, King Mongkut's Institute of Technology Ladkrabang, Bangkok, Thailand
Ekkaratadsawinnawanawa@gmail.com, boontee.kr@kmitl.ac.th

Abstract. This paper experiment on various concepts in performing mutation to lessen trap in a local optima problem of Particle swarm optimization (PSO). The first concept is when to perform mutation. The earlier mutation favors exploration more than exploitation and usually leads to slow convergence, while the late mutation tends to have opposite characteristics. The second concept is the reset of a known best position (GBEST) when trapping in local optima. The reset reduces the chance of trapping in the same local optima but may lead to slower convergence. On the other hand, mutations without reset best position exploit previous knowledge and converge faster if the GBEST closes to optima. The performances of each concept are compared using 27 benchmark test functions. The results are mixing, but the early mutation without reset GBEST perform better in many of test function.

Keywords: Local optima · Mutation · Particle Swarm Optimization

1 Introduction

PSO is famous in many optimization problems especially in continuous search space due to its fast convergence and good optimal results [1]. PSO is simple to implement and iteratively locate better solutions through the changing position of the particle (candidate solution) moving toward the best-known position. Its main drawback is trapping in local optima. There are many PSO variants try to solve the trapping problem. The main idea is trying to avoid trapping by wider global search such as multiple swarms [2, 5], perturbing of particle position [3]. The other approach is to restart swarm in the new mutation position from the trapping position [4, 6].

The first approach, that wider the search by perturbing normal PSO moving toward the best position, trade of exploration to exploitation. It causes the global search wider at the cost of fast convergence (local search). On the contrary, the second approach allows PSO to trap in local optima and restart swam in a new mutation position. This approach has fast convergence and favors local search toward the best position than global search. Also, PSO may trap again in the same trapping position. The difference between these two approaches is when a mutation occurs. The early one during the PSO iteratively locates a better position before trapping and the later one after trapping. On the extreme, muting too fast will prevent convergence and lead to poor local

P. Meesad and S. Sodsee (Eds.): IC^{2}IT 2020, AISC 1149, pp. 149–158, 2020.
https://doi.org/10.1007/978-3-030-44044-2_15

optima. Muting to slow will limit exploration and also wasting search time in the same local optima.

The other variation is the use of the best position, changing particle position without changing the best position will draw a particle back to the known best position unless there is a new best position. To experiment on these mutation variations, the same core variant of PSO [6] is used and each addition variant is implemented and tested on benchmark functions.

This paper is organized into five sections. The first one is the introduction followed by previous works, the detail description of the algorithm in the experiment on mutation variations, experiment results and conclusion.

2 Previous Works

2.1 Particle Swarm Optimization (PSO)

PSO [1] locate an optimal solution in continuous search space by iteratively changing particle (solution) toward the best-known position (Gbest) among all particle in the swarm. A Particle also moves toward its best-known position (Pbest) and its previous momentum, as shown in Eqs. (1), (2). If any particle locates a better position, its best position Pbest is updated and if the position is better than Gbest, Gbest is updated to the new position. By iteratively changing position toward best position and update new best position, PSO keeps finding new optima until no new best position found and PSO trap in the current optima.

$$V_i = (\omega V_i) + C_1 R_1 (GB - X_i) + C_2 R_2 (PB_i - X_i) \quad (1)$$

$$X_i = V_i + X_i \quad (2)$$

i is an index of the ith particle, V_i is the velocity of particle i^{th}, X_i is a position of particle i^{th}, PB_i is Pbest of particle i^{th}, GB is Gbest, ω is inertia weight, C_1 is social parameter and C_2 is cognitive and R_1, R_2 are random weights from 0 to 1.

2.2 Enhance Particle's Exploration of Particle Swarm Optimization with Individual Particle Mutation (IMPSO) [6]

This research alters PSO by mutating particle position through muting its Pbest Eq. (3) in 10% of the dimensions. A particle needs mutation when its Pbest stops improving for some consecutive amount of iteration. In comparison to muting all particles when Gbest stops improving, the monitoring of Pbest will occur faster and not all particles mutate at the same time. Since the mutated particle mutes through its Pbest instead of muting its position directly. No need to reset Gbest since Pbest will pull the particle from its normal path while other particles still moving toward current Gbest. This mutation will not disturbance the normal convergence of PSO. The Pseudo Code of the algorithm is shown below.

$$PBest_i = \pm(PBest_i \times (0.9 + 0.1 \times rand(0,1))) \quad (3)$$

```
IMPSO Pseudo Code

#Particle Swarm Initialization
Initial parameter V, X and Counter to All particle

Define Threshold = 30
While(Iteration < MaxIteration) :

  PSO Move() #According to Equation (1) and (2)
  for i = 0 to SwarmSize: #Pbest and Gbest Update

    If(Fitness(X(i)) < Fitness(Pbest(i))): #Pbest's Improved
      Pbest(i) = X(i) # Pbest Update
      Counter(i) = 0 # Reset Counter
    Else:# Pbest's not Improved
      Counter(i) += 1

    If (Counter(i) > Threshold):# Conduct the Mutation process
      Counter = 0
      Mutation(Mutate only 10% of Pbest(i)'s Dimension with Equation (3))
    If(Fitness(Pbest(i)) < Fitness(Gbest)) :
      Gbest = Pbest(i)

   If (Condition's satisfied):
     Break
   Iteration += 1
```

3 Proposed Mutation Variations

The first variation, IMPSO1, is the late mutation where all particles are mutated through their Pbest using Eq. (3) when Gbest stops improving 100 iterations consecutively. Gbest is still at the trap position and all particles might be drawn back to Gbest again unless there is a new Gbest before all particle trap again.

The second variation, IMPSO2 is also the late mutation but with reset Gbest to the best Pbest after mutation. Since Gbest is reset, there are more searches around new mutation Pbest, and the swarm has less chance to trap in the same position.

The last variation, IMPSO3, is the same as IMPSO2 with the addition of mutating Gbest further with the Eq. (4) in 30% of the dimensions. Since Gbest of IMPSO2 is further mutate to drawn the swarm further away from the previous position.

$$GBest = \pm(GBest \times (0.9 + 0.1 \times rand(0, 1))) \tag{4}$$

```
The Pseudo Code of IMPSO1-3 is as follow:

#Particle Swarm Initialization
Initial parameter V, X to All particle

Define Threshold = 100
Define Counter = 0
While(Iteration < MaxIteration) :

   PSO Move() #According to Equation (1) and (2)
   for i = 0 to SwarmSize: #Pbest and Gbest Update

      If(Fitness(X(i)) < Fitness(Pbest(i))): #Pbest's Improved
       Pbest(i) = X(i) # PBest Update
      If(Fitness(X(i)) < Fitness(Gbest(i))):# Gbest's Improved
       Gbest(i) = X(i) # GBest Update
       Counter = 0
      Else:
       Counter += 1
     If (Counter > = Threshold):# Conduct the Mutation process
      Counter = 0
      Reset Gbest Fitness #IMPSO2&3
      for i = 0 to SwarmSize:
       Mutation(Mutate only 10% of Pbest(i)'s Dimension with Equation (3))
       If(Fitness(Pbest(i)) < Fitness(Gbest)) :
         Gbest = Pbest(i)
      Mutation(Mutate only 30% of Gbest(i)'s Dimension with Equation (4)) #IMPSO3 Only
    If (Condition's satisfied):
    Break
    Iteration += 1
```

Figure 1 shows the position of the first particle of IMPSO in 90 dimensions, where each color represents a different dimension of the first particle. The optimal point for all dimensions of Schwefel is at 420.968746. At first, the position in all 90 dimensions scatters around the search space (−512, 512). In 2400 iteration, the position converges to around 400, 200, −100, −300 in some dimensions with fitness closed to 5000. The purple color vertical line indicates the iteration that the mutation of the first particle occur, as shown in Figs. 1, 2.

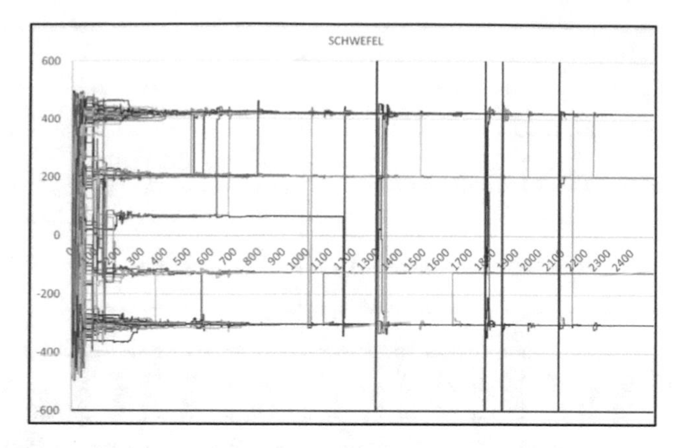

Fig. 1. IMPSO first particle position at the beginning 2500 iteration.

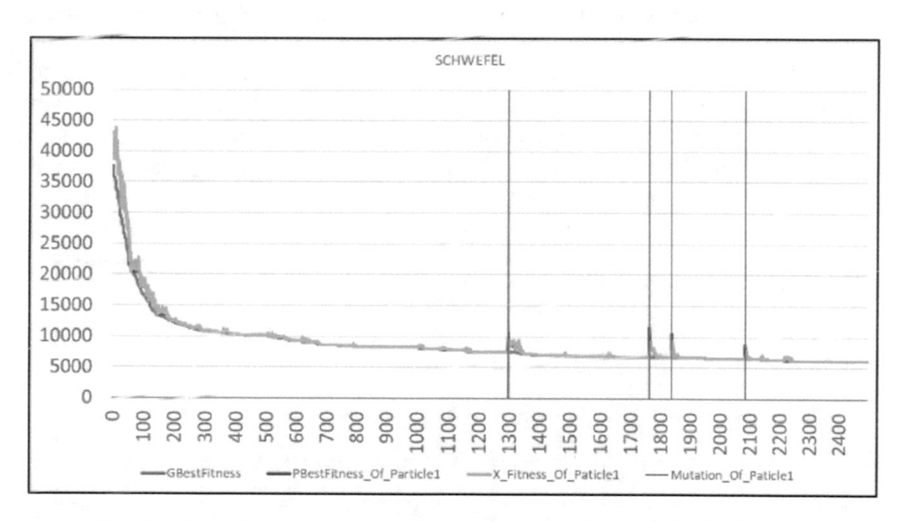

Fig. 2. IMPSO first particle fitness value at the beginning 2500 iteration.

Figure 2 shows the convergence of fitness values (y-axis) when iterations (x-axis) increase. It is magnified in Fig. 3, where the first particle1 mutation starts at iteration 1298. Before iteration 1298, this particle trapped and its Pbest unchanged. At this point, GBest, PBest and particle position (blue, red, and green color) are in the same position since their fitness value equal. Pbest (red) was muted first and the fitness value of Pbest increased (worsen) then particle (x position in green) change toward Pbest and getting worst than Pbest. Later iterations the particle position change to better fitness again drawn by Gbest (blue). Whenever particle fitness value is better than Pbest, Pbest is updated. In Fig. 3, the red and green lines do not cross blue line indicate that Gbest improved by other particles.

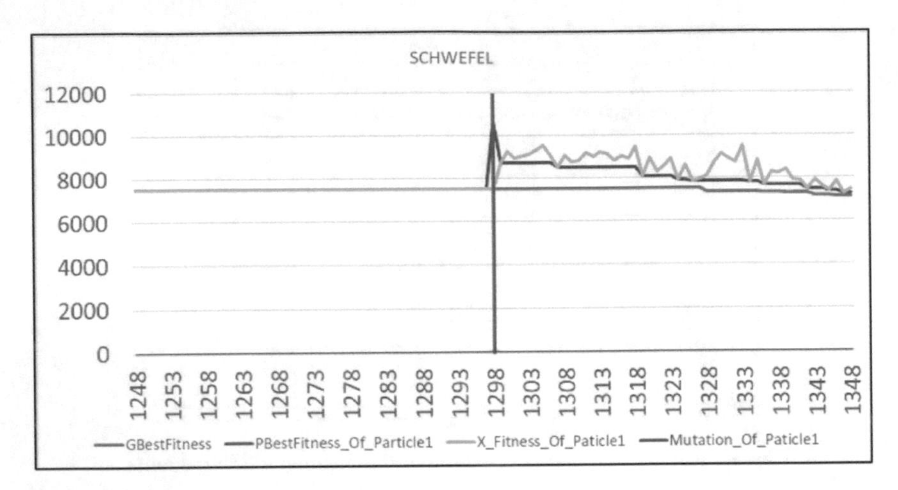

Fig. 3. IMPSO the first particle fitness and its Pbest and Gbest during 1248–1348 iteration.

Figure 4 shows that IMPSO starts mutation very fast since the threshold is set to 30. The first iteration that has mutation is at iteration 33, 83, and 111 with one particle mutation. This particle has mutation much faster than particle1 that starts its first mutation at iteration 1298, as shown in Figs. 1, 2 and 3. So not all particle mute at the same rate some slower and some faster.

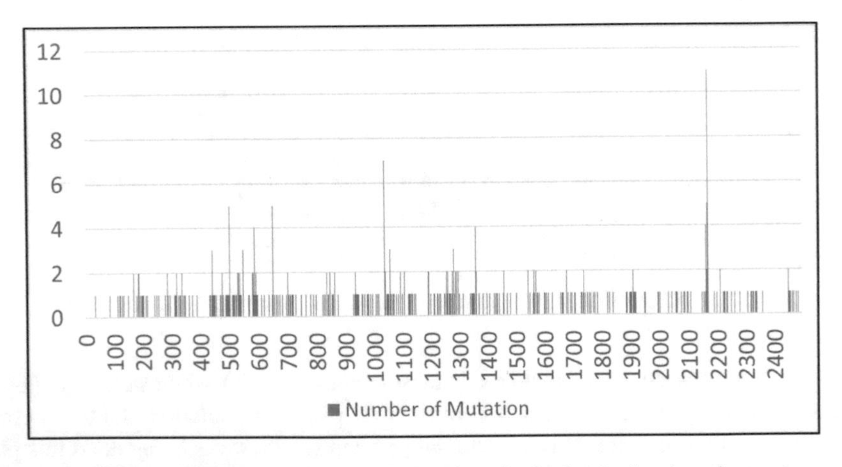

Fig. 4. IMPSO the amount of mutations in the beginning iteration.

The same kind of graphs for IMPSO1 are shown below. Since the threshold is 100 and Gbest is used instead of Pbest, so mutation occurs much slower and the swarm is trapping in local optima longer. And when performing a mutation, all particles mutation at the same trap iteration instead of individual particle mutation as in IMPSO. The fitness converges much slower as shown in Fig. 6, But this for only one sampling, overall IMPSO1 is much faster for this function, as will be seen in an average of 10 runs in the experiment results section (Table 3).

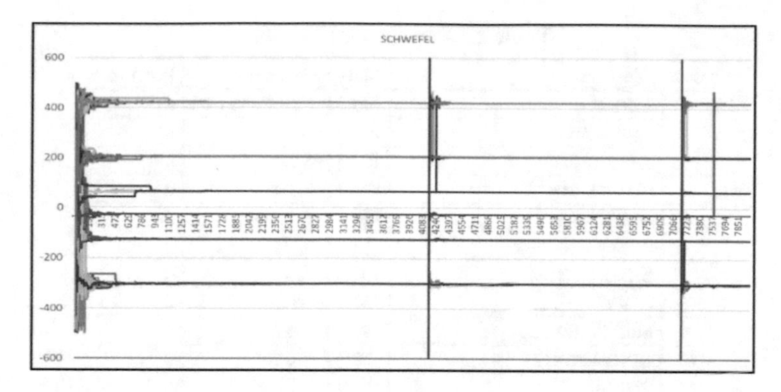

Fig. 5. IMPSO the first particle position at the beginning 8000 iteration.

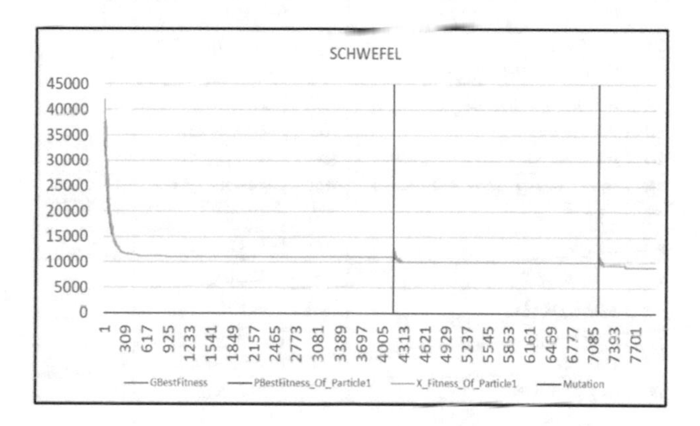

Fig. 6. IMPSO1 the first particle fitness value at the beginning 8000 iteration.

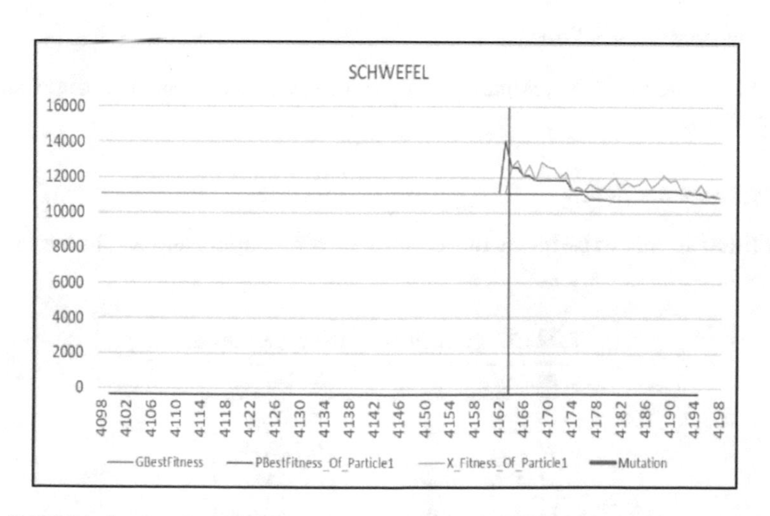

Fig. 7. IMPSO1 the first Particle fitness and its Pbest and Gbest during 4098–4198 iteration.

Table 1. Benchmark Functions.

No.	Equations	Dim	Search space	Best fitness
1	ACKLEY	100	[−32.768, 32.768]	0
2	GRIEWANK	100	[−300, 300]	0
3	RASTRIGIN	100	[−5.12, 5.12]	0
4	ROSENBROCK	100	[−2.048, 2.048]	0
5	SCHWEFEL	100	[−500, 500]	0
6	COSINEMIXTURE	100	[−1, 1]	0
7	EXPONENTIAL	100	[−1, 1]	0
8	LEVY	100	[−10, 10]	0
9	MICHALEWICZ	10	[−0, Pi]	0
10	DIXON-PRICE	10	[−10, 10]	0
11	STEP	100	[−5.12, 5.12]	0
12	SCHAFFER	2	[−100, 100]	0
13	HOLDER	2	[−10, 10]	0
14	BEALE	2	[−4.5, 4.5]	0
15	SHUBERT	2	[−10, 10]	0
16	GOLDSSTEIN-PRICE	2	[−2, 2]	0
17	SIX-HUMPCAMEL	2	[−2, 2]	0
18	SPHERE	100	[−5.12, 5.12]	0
19	PARALLELHYPER-ELLIPSOID	100	[−5.12, 5.12]	0
20	ROTATEDHYPER-ELLIPSOID	100	[−65.536, 65.536]	0
21	CIGAR	100	[−10, 10]	0
22	BROWN	100	[−1, 4]	0
23	MULTIMOD	100	[−10, 10]	0
24	ZAKHAROV	100	[−5, 10]	0
25	TRID	10	[−100, 100]	0
26	EASOM	2	[−100, 100]	0
27	EGGHOLDER	2	[−512, 512]	0

4 Experiments

4.1 Benchmark Functions

Table 1 is the list of 27 Benchmark test functions with the number of dimensions and their search space.

4.2 Experiments Setup

All algorithm in this experiment run under the same parameters as shown in Table 2 Below.

Table 2. Parameters in the experiments.

Parameter	Value
C1, C2	*1.496180*
W	*0.729844*
MAX Iteration	*90,000,000*
Size of swarm	90 particles
Number of experiment	10

4.3 Experiments Results

From the results shown in Table 3, IMPSO miss finding optima in two functions (25 and 27) while IMPSO1 misses four functions (4, 8, 25 and 27) and IMPSO2-3 miss three functions (4, 5 and 27). But IMPSO locates optima significant faster in 14 functions with green character and much slower in 2 functions (function 5 and 25). In function 25, IMPSO can't locate optima while IMPSO2-3 can locate optima very fast for this function. Overall, earlier mutation and exploitation of the best known of trap position seem to converge faster and find more optimal but not in all test functions.

Table 3. Benchmark functions optimization results.

No.	IMPSO		IMPSO(1)		IMPSO(2)		IMPSO(3)	
	Evaluation call	Best fitness value	Evaluation call	Best fitness value	Evaluation call	Best fitness value	Evaluation call	Best fitness value
1	**1,968,821**	0	5,835,952	0	5,354,479	0	5,147,368	0
2	354,382	0	392,003	0	540,235	0	602,856	0
3	**1,034,214**	0	4,196,737	0	5,863,960	0	7,612,487	0
4	**1,948,092**	0	90,889,876	6.53102E-26	90,009,370	3.64602E-28	90,009,211	3.30089E-28
5	24,631,717	0	**17,603,758**	0	90,030,943	2,297.414964	90,029,896	2579.693267
6	**348,837**	0	695,917	0	683,749	0	664,145	0
7	**675,615**	0	1,513,783	0	1,211,554	0	1,259,519	0
8	**1,545,030**	0	82,072,639	5.69515E-27	18,391,717	0	18,838,485	0
9	**65,884**	0	121,411	0	134,443	0	174,855	0
10	9,039	0	11,755	0	9,406	0	13,006	0
11	100,542	0	133,795	0	143,110	0	137,713	0
12	23,787	0	28,090	0	35,173	0	23,203	0
13	18,349	0	13,177	0	13,096	0	11,944	0
14	32,650	0	31,834	0	32,248	0	31,402	0
15	6,312	0	6,211	0	6,490	0	6,751	0
16	11,883	0	11,908	0	11,917	0	11,548	0
17	2,326	0	2,350	0	2,521	0	2,287	0
18	**5,312,789**	0	8,287,615	0	8,342,965	0	8,268,861	0
19	**5,357,988**	0	8,299,468	0	8,337,871	0	8,393,348	0
20	**5,264,152**	0	8,086,366	0	8,234,425	0	8,176,079	0
21	**5,345,748**	0	8,391,610	0	8,339,671	0	8,218,263	0
22	**5,306,954**	0	8,273,962	0	8,357,446	0	8,334,849	0
23	**13,923,337**	0	26,473,609	0	59,244,076	0	39,884,158	0
24	**26,918,558**	0	84,186,127	0	84,636,262	0	84,623,883	0
25	90,737,162	3.78956E-12	81,832,699	4.63842E-12	**312,157**	0	**192,145**	0
26	17,653	0	17,281	0	17,659	0	18,019	0
27	38,121,753	11.297052	48,622,978	37.10177858	59,487,913	37.37399082	53,387,667	26.29647488

5 Conclusion

This paper is comparing the mutation of particle position before trapping (IMPSO) and after trapping (IMPSO1-3). Also, the best-known position is reset in IMPSO2-3. From the test results, the mutation before trapping and no reset of the best-known position locate optima a little better (only 2/27 miss, while the other have 4 and 3 miss) and significantly faster in almost half (14) of the test functions. But on Schwefel function mutation after trapping IMPSO1 is much faster and also the Trid function IMPSO2-3 can locate optima while IMPSO can't. The experiment is controlled by using the same core algorithm and the same parameter with the same kind of mutation; the results are varied depending on the test function and IMPSO tends to locate optima faster.

References

1. Kennedy, J., Eberhart, R.C.: Particle swarm optimization. In: IEEE International Conference on Neural Networks, pp. 1942–1948 (1995)
2. Poempool, L., Kruatrachue, B., Siriboon, K.: Combine multi particle swarm in supporting trapping in local optima. In: Applied Sciences and Technology, pp. 416–419 (2018)
3. Xinchao, Z.: A perturbed particle swarm algorithm for numerical optimization. Appl. Soft Comput. **10**(1), 119–124 (2009). https://doi.org/10.1016/j.asoc.2009.06.010
4. Cheypoca, V., Siriboon, K., Kruatrachue, B.: The use of Global Best position in rerun of particle swarm optimization. In: Applied Sciences and Technology, pp. 610–613 (2018)
5. Cheung, N.J., Ding, X.-M., Shen, H.-B.: OptiFel: a convergent heterogeneous particle sarm optimization algorithm for takagi-sugeno fuzzy modeling. IEEE Trans. Fuzzy Syst. (2013). https://doi.org/10.1109/TFUZZ.2013.2278972
6. Adsawinnawanawa, E., Kruatrachue, B., Siriboon, K.: Enhance particle's exploration of particle swarm optimization with individual particle mutation. In: The 7th International Electrical Engineering Congress, Cha-am, Thailand, 6–8 March 2019, pp. CIT49–CIT52 (2019)

Intruder Detection by Using Faster R-CNN in Power Substation

Krit Srijakkot, Isoon Kanjanasurat, Nuttakan Wiriyakrieng, Mayulee Lartwatechakul, and Chawalit Benjangkaprasert(✉)

Department of Computer Engineering, Faculty of Engineering, King Mongkut's Institute of Technology Ladkrabang, Bangkok, Thailand
krit.srij@gmail.com, pe_win99@hotmail.com, nuttheguitar@gmail.com, {mayulee.la, chawalit.be}@kmitl.ac.th

Abstract. This paper presents the intruder detection by using the Faster R-CNN model and administrator system for the power substation in Khon Kaen substation 4 of the Electricity Generating Authority of Thailand (EGAT). There are two processes of intruder detection-detecting the intruder and sending a notification to the system administrator of EGAT through Line application. The Faster R-CNN model of intruder detection was trained and tested by using the Open Image Dataset and our dataset. We collected our dataset of 1,500 images from a different condition from the real environment. There are two conditions, including distance and light intensity. Our system used a high-performance computer by using GPU: Nvidia Titan RTX 24 GB to support the object detection system from using five cameras at the same time. The performance of intruder detection achieved by greater than 95%.

Keywords: Faster R-CNN · Intruder detection · Object detection

1 Introduction

By security and electric stability reasons, the power substation area must be the restrict area. The problem occurs when the intruder comes into the restrict area as the system administrator misses monitoring that might make an unexpected situation. This intruder detection system would cover this unexpected situation and record them to investigate. It uses the surveillance cameras for monitoring intruders in the power substation area. Although surveillance can detect intruders and replay the previous situations that recorded in the video, it cannot monitor the intruders all time and prevent the unexpected situation.

We have studied the previous researches work that using the detection models to detect and classify the object. Critical points of detection are accuracy and object-classification. In this work, we choose Faster Region Convolution Neural Network (Faster R-CNN) [1] method, then including some techniques to improve the performance of the model to detect intruders in a restricted area.

The detection based on Faster R-CNN has been developed from the Convolution Neural Network (CNN) [2]. In converting the picture to the feature map process, CNN

P. Meesad and S. Sodsee (Eds.): IC^{2}IT 2020, AISC 1149, pp. 159–167, 2020.
https://doi.org/10.1007/978-3-030-44044-2_16

takes a long time for data processing. Besides that, CNN's performance does not adapt to use in real-time detection. The evolution of model detection from CNN to Region Convolution Neural Network (R-CNN) needs a smaller number of inputs [3]. Although R-CNN has the ability to detect and use Support Vector Machine (SVM) [4] to classify types of objects, the efficiency of detection is not enough for the real situation. R-CNN will proceed selective search before feeding them into the CNN that used long time-consuming and got about two thousand feature maps. The new process can skip the selective search processes to feed a single input image to CNN, so the Fast R-CNN, which resulted from the CNN, would be only one feature map from one input image [5]. Gavrilescu et al. [6] used the Faster R-CNN model to detect traffic indicators and evaluated his model in different light-condition in each period of the day both day and night time to test its application performance. This study was being trained and tested at a speed of pictures of 15 fps on a set of a dataset containing 3,000 images. It found that they could adapt the Faster R-CNN for real-life situations, but could not indicate how far the object was. Liu et al. [7] showed object detection based on Faster R-CNN and used a different dataset to represent the performance of detection and classification using Faster R-CNN. This study classified objects such as humans, cats, cars, etc. However, this paper only concerned about detection efficiency but did not involve the condition of detection distance and the brightness, which is important in determining the performance of detection.

Focusing on the human-object model, Hsu et al. [8] identified human by using Fast R-CNN. This paper was to detect human-body as each separate-part and set the highest score when human was detected. Although the score of detection was better than other methods, the paper did not include the distance and brightness conditions, while these two factors affected the detected model. Wang et al. [9] created the scheme for moving object detection for the smart substation. This testing had done about real-time detection and quality of the video but not included two main factors, such as distance and light condition. Mohan and Resmi [11] puts the background subtraction technique in image processing. The detect-moving object and the background were compared. In comparison with the results from no background subtraction, the background subtraction technique was better. However, the condition such as the distance or brightness of the area was not concerned.

This research aims to upgrade the performance of the surveillance camera detecting system of the EGAT power substation to reduce monitoring tasks by adding necessary information to the system. Then, import the output from surveillance cameras to the system when the human invader in the restrict area has detected so that the system makes a notice to the system administrator by sending the notification via Line application. By this method, the smaller number of patrol people is also benefiting.

In this work, the system consists of two main processes that require Faster R-CNN and Line notification system using Line application programming interface (API). There is the detecting of the intruders in the power substation restricted area and the notifying the system administrator when the intruders are found.

2 Methodology

2.1 Faster Region Convolution Neural Network (Faster R-CNN)

Faster Region Convolution Neural Network or Faster R-CNN [1] was developed from the previous model, Convolution Neural Network (CNN) [2] which was adapted for increase performance of detection and reduce time in convolution process. CNN was the basic method for changing the picture to binary, but the occurred convolution required the performance and time-consuming in the process while Faster R-CNN could reduce the bottleneck in the CNN. We already calculated images with CNN instead of running a separate discriminating exploration algorithm on the feature map to classify the region proposals such as Fast R-CNN. Faster R-CNN reused the same CNN results to predict region proposals. Then, there were reshaped using Region of Interest (RoI) pooling layer which was used to classify the images within the proposed region and prognosticate the offset values for the bounding boxes (see Fig. 1).

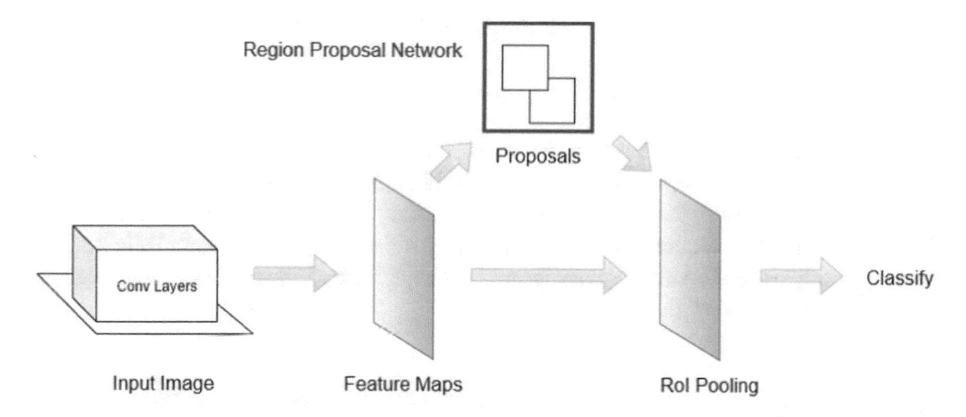

Fig. 1. Block diagram of Faster R-CNN.

2.2 Overall Process

The overall process of the system, it consists of two main subsystems: object detection and line notification (see Fig. 2).

2.3 Object Detection System

Object detection system using the DOCKER platform manages the system by docker-compose and connect the system with the database. The primary process used TensorFlow for running model Faster R-CNN to detect and classify the intruders. The object detection system would request the video from surveillance cameras by using Faster R-CNN detection and classification. Then, the system collects the information and sends it to the database and sends it to Line application for notifying. This process was the main process shown in the middle (see Fig. 2).

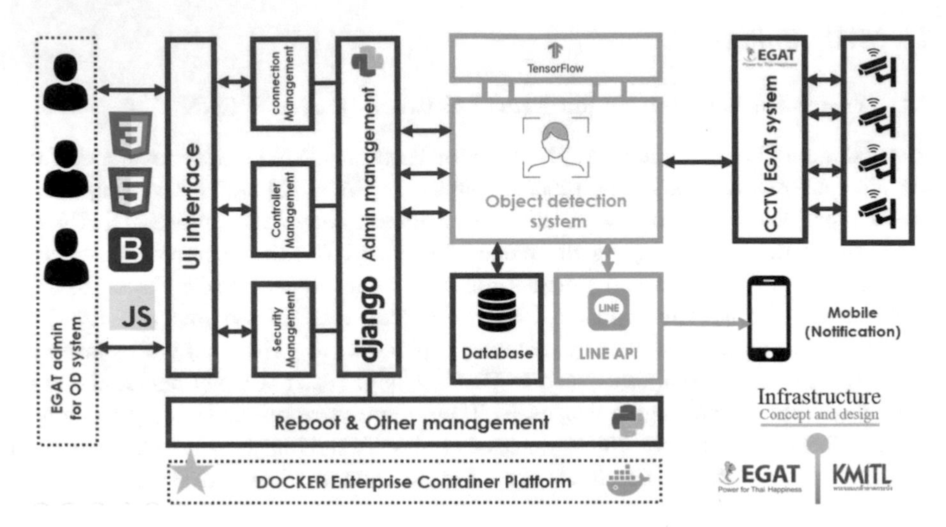

Fig. 2. Overall process of object detection and line notification system.

2.4 Line Notification System

This system uses Line notify to report when the problem occurs. Line Bot in notification system would send the details about type, image and the position of the intruder. So, the system administrator would make it easier to manage when an unexpected situation occurs. The sequence of Line notifies by object detection system followed this block diagram since the intruder was detected. The picture of the intruder would be sent out via Line application (see Fig. 3).

Fig. 3. Line notifies diagram.

2.5 Mean Average Precision

Mean Average Precision or mAP [10, 12] represented the performance value of detection model by comparing train set and test set. mAP would calculate from all average precision and average all of them. The mean average precision is defined as an Eq. (1)

$$mAP = \frac{1}{N}\sum_{i=1}^{N} AP_i \tag{1}$$

Where, AP is the average precision and N is the number of classes.

The AP [12] could be precision curve, and is defined as an Eq. (2)

$$AP = \frac{1}{M_p} \sum_{i=1}^{M_p} Pr(i) \tag{2}$$

Where, Pr is the precision at each r level, which showing the maximum precision and M_p is the number of sample-image.

3 Experiments and Results

For implementation and collection of the results from the system, we use a high-performance computing computer by using GPU: Nvidia Titan RTX 24 GB to support object detection system five cameras at the same time. Faster R-CNN, the model which is being used in our system, was pre-trained by using Open Images Dataset (OID) [13]. And for the appropriate performance, we collected the datasets from the real environment mix up with datasets from the OID. In our training process, we split 70% of data to be a training set and 30% of data to be a test set for the model evaluation process. We collected the intruder in the real environment images from different aspects of the camera and set up a different position and distance of an intruder before saving the image into the dataset.

After training the model-built object detection system, we implement the object detection system with the Line notification API service system to notify when the system detects the intruder. The Docker service becomes our platform to manage communication between both systems. When starting the system, the object detection system must send only an interesting type of intruder to the Line notification API service system to confirm the performance of our system.

In our experiment, we split the dataset into two groups. The first group contains 2,500 images of persons from the Open Image Dataset and 2,500 images of the person who intrudes from the real environment. The second group includes 5,000 images of the intruder only from the real environment. The mAP score used as a model testing result performance from both groups of the dataset. Our test set, which contains 1,500 images and collected in a different condition from the real environment, was split equally. The situation which chooses to test the model performance is inside the building and outside the building light intensity, not only the daytime but also in the nighttime.

The bounding box which detects an intruder at night inside the building, daytime and night outside building which come from different inspection of cameras in the power substation (see Figs. 4, 5, 6) respectively.

Fig. 4. Intruder detection inside building.

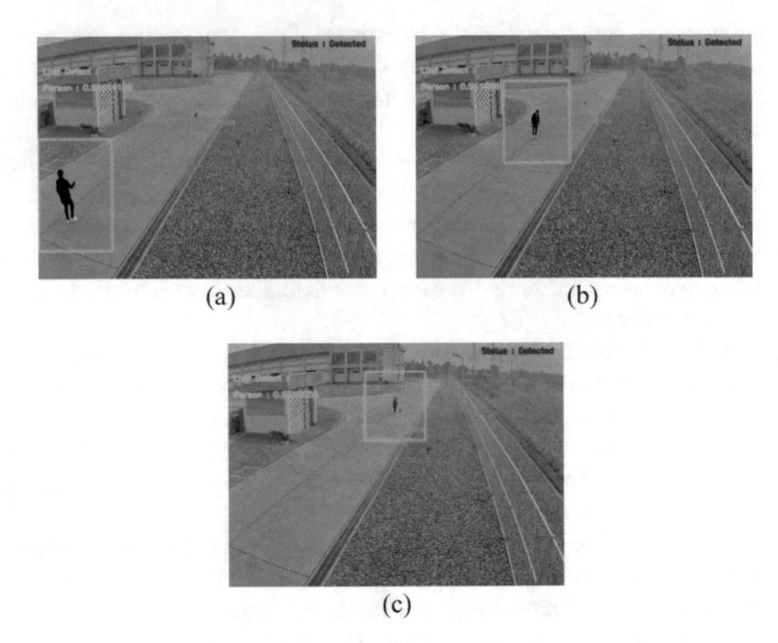

Fig. 5. Intruder detection outside building during daytime: (a) 5 m, (b) 10 m and (c) 20 m, respectively.

Type of intruder, camera, date & time and shows picture of intruder in Line application. This information would alarm to system administrator and then system administrator prevents unexpected situation before occurring (see Fig. 7).

The result from Table 1 showed that an object detection model in which training from the real environment image and Open Image Dataset have a better mAP score in all conditions. When looking at the inside building image testing, the model with Open

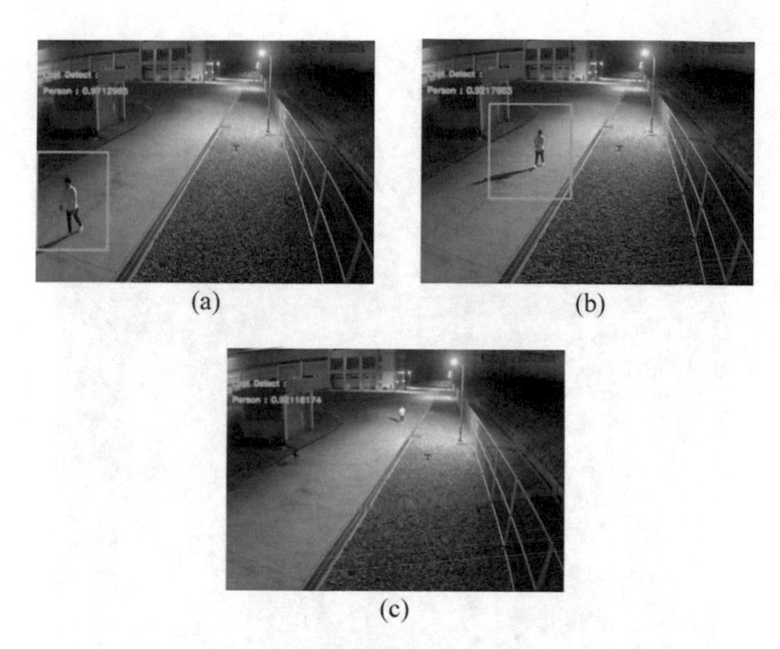

Fig. 6. Intruder detection outside building during nighttime: (a) 5 m, (b) 10 m and (c) 20 m, respectively.

Table 1. The average of mAP score from a different model which training from two groups of the dataset.

Environment condition	Dataset without Open Image Dataset	Dataset with Open Image Dataset
mAP inside building	0.901	0.965
mAP outside building daytime	0.987	1.000
mAP outside building nighttime	0.896	1.000

Image Dataset has the lowest mAP score when comparing with outside building in both daytime and nighttime. But outside the building condition at the nighttime is the worst-case predictive performance of the model that training without Open Image Dataset.

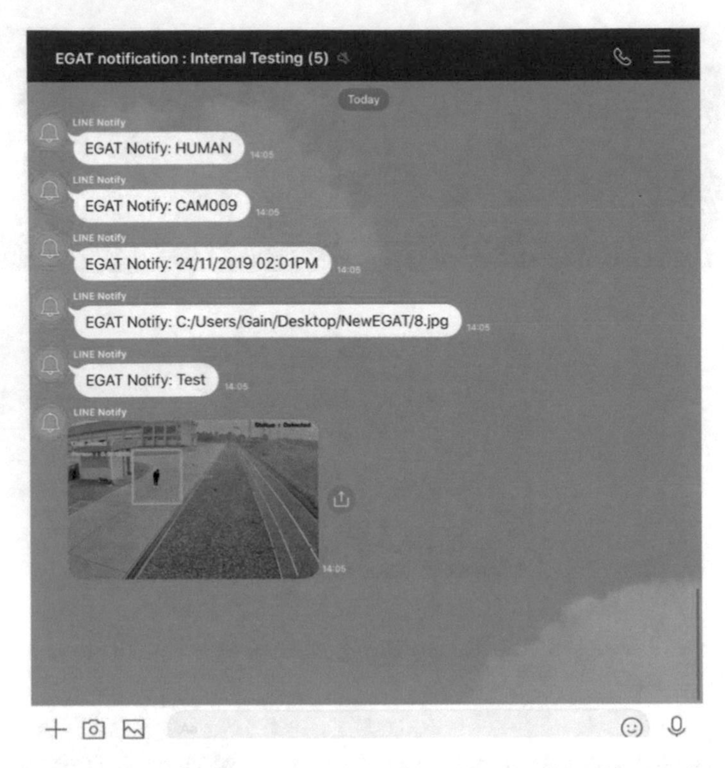

Fig. 7. Example notification when intruder in restrict area from Line application.

4 Conclusion

This paper presented the system to detect an intruder who invades the restricts area in the power substation. The results from testing were very good to detect humans and send messages to notify when the intruder is invading to the restrict area. In Fig. 4, the results inside the building were better than outside the building in Fig. 5. Because the light condition in the building makes an appropriate condition for detecting the same as outside the building in the daytime, this light condition made the model perform the highest performance of the detection model. But at night, the light state makes the image quality from surveillance cameras worse than daytime and it reduces detecting performance.

From the above experiment, to use a model in a real environment, not only the dataset from the real environment can improve a predictive score, but the open-source image, for instance: OID also helps to improve a model performance when doing a training process.

Acknowledgments. This research was supported by The Electricity Generating Authority of Thailand (EGAT). We thank our colleagues from King Mongkut's Institute of Technology Ladkrabang (KMITL) who provided insight and expertise that greatly assisted the research.

References

1. Zhang, H., Du, Y., Ning, S., Zhang, Y., Yang, S., Du, C.: Pedestrain detection method based on faster R-CNN. In: International Conference on Computational Intelligence and Security, pp. 427–430 (2017)
2. Kannojia, S.P., Jaiswal, G.: Ensemble of hybrid CNN-ELM model for image classification. In: International Conference on Signal Processing and Integrated Networks, pp. 538–541 (2018)
3. Zhang, W., Wang, S., Thachan, S., Chen, J., Qian, Y.: Deconv R-CNN for small object detection on remote sensing images. In: IEEE International Geoscience and Remote Sensing Symposium, pp. 2483–2486 (2018)
4. Wang, X.Z., Lu, S.X.: Improved fuzzy multicategory support vector machines classifier. In: Fifth International Conference on Machine Learning and Cybernetics, pp. 3585–3589 (2006)
5. Shih, C.H., Chung, L.H., Cheng H.C.: Vehicle detection using simplified fast R-CNN. In: International Workshop on Advanced Image Technology, pp. 1–3 (2018)
6. Gavrilescu, R., Zet, C., Fosalau, C., Skoczylas, M., Cotovanu, D.: Faster R-CNN: an approach to real-time object detection. In: International Conference and Exposition on Electrical and Power Engineering, pp. 165–168 (2018)
7. Liu, B., Zhao, W., Sun, Q.: Study of object detection based on faster R-CNN. In: Chinese Automation Congress, pp. 6233–6236 (2017)
8. Hsu, S.C., Wang, Y.W., Huang, C.L.: Human object identification for human-robot interaction by using fast R-CNN. In: Second IEEE International Conference on Robotic Computing, pp. 201–204 (2018)
9. Wang, Y., Zhang, J., Zhu, L., Sun, Z., Lu, J.: A moving object detection scheme based on video surveillance for smart substation. In: 14th IEEE International Conference on Signal Processing, pp. 500–503 (2018)
10. Li, K., Huang, Z., Cheng, Y.C., Lee, C.H.: A maximal figure-of-merit learning approach to maximizing mean average precision with deep neural network based classifiers. In: IEEE International Conference on Acoustic, Speech and Signal Processing, pp. 4503–4507 (2014)
11. Mohan, A.S., Resmi, R.: Video image processing for moving object detection and segmentation using background subtraction. In: First International Conference on Computational Systems and Communications, pp. 288–292 (2014)
12. Kim, I., Lee, C.H.: Optimization of average precision with maximal figure-of-merit learning. In: IEEE International Workshop on Machine Learning for Signal Processing, pp. 1–6 (2011)
13. Niitani, Y., Akiba, T., Kerola, T., Ogawa, T., Sano, S., Suzuki, S.: Sampling techniques for large-scale object detection from sparsely annotated objects. In: Computer Vision and Pattern Recognition, pp. 6510–6518 (2019)

Super-Resolution Image Generation from Enlarged Image Based on Interpolation Technique

Athaporn Kingboo[1], Maleerat Maliyaem[1](✉), and Gerald Quirchmayr[2]

[1] Faculty of Information Technology, King Mongkut's University of Technology North Bangkok, Bangkok, Thailand
atha2556@hotmail.com, maleerat.s@it.kmutnb.ac.th
[2] Multimedia Information Systems, Faculty of Computer Science, University of Vienna, Vienna, Austria
gerald.quirchmayr@univie.ac.at

Abstract. This research proposed a Super-Resolution Image Generation (SRG) from enlarged image based on bicubic interpolation technique in order to reconstructs a higher-resolution image. This technique uses neighboring pixels to calculate a value for adjusting an appropriate coefficient of a new pixel that given higher resolution. SRG technique is developed based on popular pixel estimation called Bicubic technique which widely used for image resolution adjustment. The performance is evaluated based on PSNR and SSIM measurement, the results showed better overall reconstruction quality in terms of resolution and sharpness.

Keywords: SRG · Bicubic · Image enhancement · Enlarge · Resize · Interpolation

1 Introduction

Digital Video Recorder (DVR) is a device that converts the analog signals from a CCTV camera to digital format. Digital images can usually be divided into two distinct categories. They are either raster (bitmap) files or vector graphics [1]. Raster images are created with pixel-based programs or captured with a camera or scanner. They are more common in general such as jpg, gif, png and are widely used on the web. Vector graphics are created with vector software and are common for images that will be applied onto a physical product. Using raster images to enlarge image in the spotlight within the image or enlarge image to be larger than the original image [2]. The extended image will contain either the same or small color pattern as shown in Fig. 1. The result after image extension was not sharp and the image is blurred without the resolution of the image [3]. Especially, the single image or multi images obtained from the low-quality recorder of Closed-Circuit Television (CCTV) or smartphone will produce an image output with not good resolution and sharpness.

P. Meesad and S. Sodsee (Eds.): IC^{2}IT 2020, AISC 1149, pp. 168–180, 2020.
https://doi.org/10.1007/978-3-030-44044-2_17

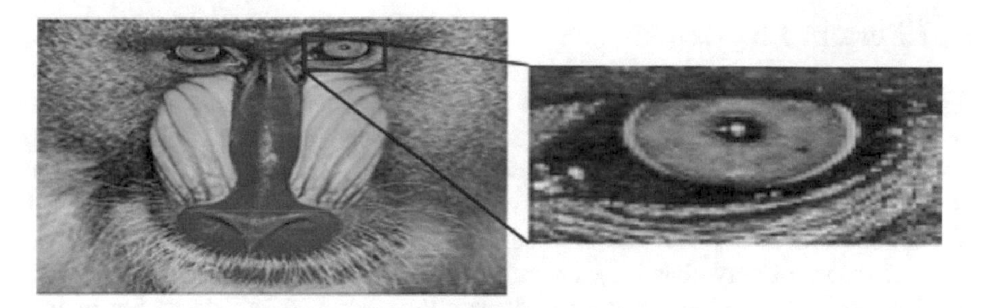

Fig. 1. Example of raster image.

This problem caused by CCTV recording system and various environments. This is an important problem if needed to enlarge image based on this input characters. Therefore, the performance of enlarged image needs to improve in terms of sharpness of the image quality. Previously, many techniques have been developed in order to improve image quality in terms of higher-resolution and sharpness from original image magnification using different methods in various aspects [4]. However, several techniques are commonly used to improve image quality such as interpolation, brightness, Filter, and Super-Resolution etc. Interpolation is a method that continuously developed and widely applied to improve the quality of image in terms of increasing resolution and clarity image that obtained from enlarged image. Super- Resolution (SR) techniques are a more effective way to improve image quality with greater resolution and clarity. Commonly used for image data processing that is still or motion. To create images and increase image resolution from multiple original images low resolution [5–7]. The SR will high frequency data that is lost or attenuated during the recording of image data of the recording device [8], which will result in more resolution images. The SR has been developed continuously. There are many methods that can be used to improve image quality, that is a single image or moving effectively [8–10]. SR is effective at improving the quality of moving images well. But the SR still cannot solve the problem of low quality still images from images with low-resolution images that are very dark or very bright, especially for still images with noise or blur [11, 12]. The image created from the SR will produce an increase in resolution but lack of sharpness. Image color is distorted. The image lacks the picture border. The image focuses on one of the most valuable colors. The image is not sharp [1, 3]. The image generated by the SR will have a larger image size than the original image, thereby draining the device space for the image storage. It also requires hardware performance for high-performance image processing. These problems, even though there have been improvements and improvements, but still cannot solve the problem of low-quality images effectively. This paper proposed the SRG technique for improving the quality of single image into spatial domain quality improvement. The Process is built from low-quality magnification images, especially images with lots of noise. An image is estimated the quality of a low image with multiple images using pixel around the point that you expected to estimate color, size 4×4 pixels. This technique will be evaluated based on Peak Signal-to-Noise Ratio (PSNR) and Structural Similarity Index (SSIM) measurement.

2 Literature Review

According the survey related to an image enhancement, there are many techniques used to improve the quality of enlarged images [2]. These techniques are different, in terms of improving image quality processing speed and reducing complexity of the technique [4, 5]. Improving the quality of images obtained from enlarging images is the easiest way to repeat for each pixel. This method discovered the serrated edges that are shaped in small shapes or very alias [13], image without resolution and clarity. However, this problem needs to be solved in order to increase the resolution and clarity for the image obtained from the extension which has documents and research related as follows.

2.1 Super-Resolution (SR)

Super-Resolution is a signal processing method that used to improve image quality in terms of resolution enhancement [7, 12]. From the literature reviews, many researchers focused on example-based single image SR topic. There are two related principles for image processing: Super-Resolution Reconstruction and Super-Resolution Restoration [14]. These two processes focus on trying to take high-frequency data that is lost or dropped during data recording. The Super-Resolution is a method of trying to create a restore image [15]. This information is typically reduced due to the higher frequency than the cut-off frequency of DVR or high-frequency data that damaged by aliasing issues. The creation of high-resolution images will learn a relationship between each low-resolution image, as shown in Fig. 2 and create a high-resolution image [5].

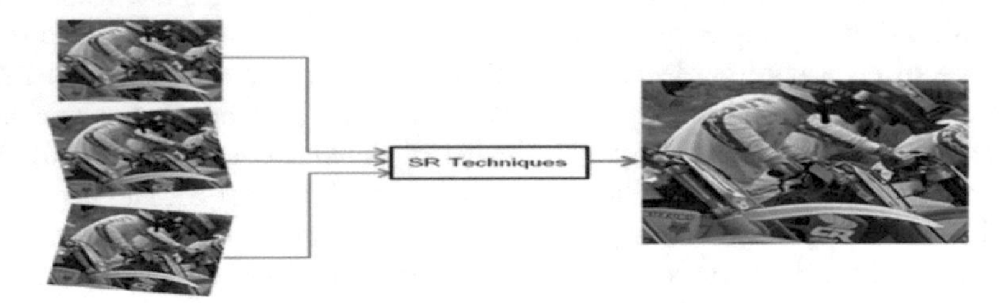

Fig. 2. A concept of Super-Resolution technique.

Therefore, the method of creating an image will eliminate blur and noise that occurs with the low-resolution image. However, there are many techniques used to create high-resolution images [1, 12]. A Super-Resolution technique is divided into three groups: interpolation-based, reconstruction-based, and learning-based [14, 16] for two types of image processing [4, 17] single and multi-images and constantly evolving [5, 18]. High-resolution and sharpness can be created from images with low resolution and sharpness but creating images from a single image cannot increase the resolution and sharpness of image particularly if a single image contained a lot of noises. However, in order to solve the problem of resolution and sharpness of increasing ages, the basic principles of SR techniques and the basic of technique Interpolation for creating a single image in the image area with different image characteristics are applied.

2.2 Interpolation

Interpolation is a method for image processing that estimate or predicts values to the pixels within the image by assigning color values to the pixels of the new image created from limited color point data samples [5, 18]. This method can be used to estimate unknown values from any point of the pixel in the image. It is a technique used for improving the quality of various images, in the resolution of the image. The important process of the technique is to interpolate the value of the pixel or the brightness value of the new pixel by adding a new resolution to the image [4, 7]. The current technique has been developed to the enhancement of the interpolation include high frequency by discrete, bi-cubic sub-band and wavelet near/dual tree-complex wavelet [5, 18]. The sample is shown in Fig. 3, input data is placed in the middle position between the pixels. Inside the image, pixels are placed anywhere within the image.

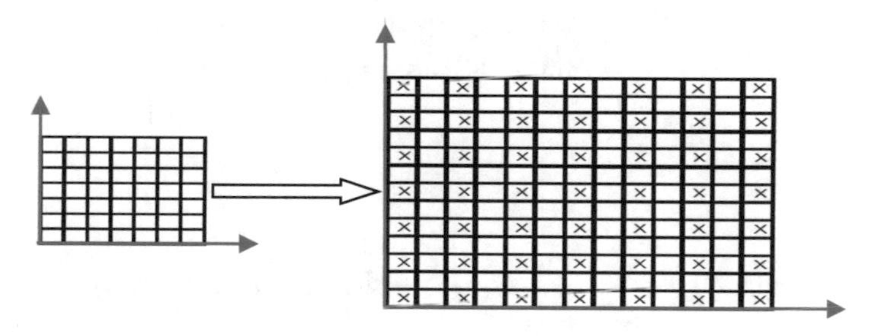

Fig. 3. From m × n to image size 2m − 1 × 2n − 1.

One problem that occurs when creating a small raster image data with that interpolation method, the result will be reduced to the size of the area that you want to create the image when placing the data in pixels. It may not be accurate that the pixel will always get the right information exactly the same default value. It works with the assumption of how the color values are dispersed in different pixels. In a related study area or may be a result of similar and have the same tendency. The general equation of interpolation from the weight of data is as follow.

$$\hat{Z}(S_0) = \sum_{i=1}^{N} \lambda_i Z(S_i) \tag{1}$$

Where

$Z(s_i)$ = The pixel value at the position
λ_i = Weight from the pixel that needs interpolation
$\hat{Z}(s_0)$ = New pixel
N = The total number of pixels.

2.3 Bicubic Interpolation

Bicubic Interpolation is an image processing technique [11, 5] constantly evolving and developing in an application of the most widely used imaging program [18]. The most common techniques are Bicubic, Bicubic Smoother, Bicubic Sharper. These techniques are techniques for increasing the resolution and sharpness of images that we enlarge images to a larger size than the original [19]. The method of this technique is about the pixel value that we are interested. The basic processing based on the interpolation of colors in the pixels around the point of interest. The surrounding pixels are interested in being called "Nearest Neighbor" [20]. By interpolation the color of the neighbor pixels, on average 16 pixels or 4 × 4 pixels. For calculating the desired pixels, the interpolation is applied in the dimension, there are two models Bicubic is One-dimension and Two-dimension [3], which has the equation follow.

One-dimension

$$u(s) = \begin{cases} (3/2)|s|3 - (5/2)|s|2 + 1 & 0 \leq |s| < 1 \\ (-\frac{1}{2})|s|3 - (\frac{5}{2})|s|2 - 4|s| - 2 & 1 \leq |s| < 2 \\ 0 & 2 < |s| \end{cases} \quad (2)$$

Two-dimension

$$g(x, y) = \sum_{i=-1}^{2} \sum_{m=-1}^{2} c_{j+i,k+m} u(distance_x) u(distance_y) \quad (3)$$

Where

g(x, y) = New pixel,
j, k = Image size (pixel)
i, m = Neighbor 4 × 4
u(s) = Distance

Based on the Bicubic Interpolation technique and the ability to Interpolation pixel values of images as well as having good filtering performance. Therefore, the bicubic interpolation technique has been introduced to improve the resolution and sharpness images from the enlarged of images generated by the SR technique.

2.4 Performance Measure

2.4.1 Peak Signal to Noise Ratio (PSNR)

PSNR is a tool to evaluate performance [21]. The most popular image quality improvement techniques calculated value is compared with the original image [17, 18]. If the PSNR value is high, it indicates the image quality that is close to the original image which is a standard used to compare the quality of the digital image. The performance of resolution technique and image filtering can be calculated from the equation follow.

$$PSNR = 10 \log_{10} \frac{b}{RMSE} \quad (4)$$

Where
$b = 255.$

2.4.2 Root Mean Square Error (RMSE)

Root Mean Square Error (RMSE) is a standard way to measure the error of a model in predicting quantitative data. RMSE can be calculated from the equation follow.

$$RMSE = \sqrt{\sum \frac{(Predict - Obs)^2}{n}} \quad (5)$$

Where
Predict = Predicted value
Obs = Observed valuc
n = Example size.

3 Research Methodology

There are various techniques increase image quality in terms of resolution and clarity. But the image that has been enlarged especially the images that have a lot of noise. These techniques are still not effective [1, 2]. At present, there are popular tcchniquc to develop and use in the expansion of imagcs. These techniques are a way to increase image quality in the resolution and sharpness of images which have different results. The proposed is a performance improvement of the Bicubic in terms of increasing resolution and sharpness. With increasing the ability to filter the noise of the image that has been expanded which has methods and step to develop the following technique.

3.1 Material and Method

The most techniques used in image processing are black-and-white, gray scale images and color images. In black-and-white images, a value is 0 or 1 pixels. Gray scale images have pixels in the range 0–255. The image is normally in color and two-dimensions in the direction of the x, y. The image used in the test is a picture. The raster has a 3D dimension according to the image processing format of the color formation red, green and blue or RGB. The image data is divided into three-dimensional processing and each dimension will be adjusted to a uint8 or 8 bit image, in each pixel of the image will be replaced with numbers that are a color values in each dimension 8 bit image data in each pixel is in the range 0–255. This will generate an image data in color format size 255, 255, 255 or about 16 million colors, making the image data very large. Therefore, to use a lot of memory for processing. Many researches in the area related to image processing, it is not popular to use color images

(a) Lena (b) Parrots (c) Bike

Fig. 4. Original images.

Table 1. Images properties.

Image	Type of file	Dimensions (Gray)	Size (KB)	Total images	Class
Lena	Bitmap	309 × 309	95.2	100	Uint8
Parrots	Tiff	256 × 256	63.7	100	Uint8
Bike	Tiff	256 × 256	64.6	100	Uint8

unit 8 or unit 16 because it consumes memory for processing and slow processing. Images are generally stored in many standard formats. These patterns affect the sharpness and quality of the image.

Due to the technique of image format, some techniques require efficiency in reducing the amount of data to be stored cause to lose some data. In the experiments, all images used in the experiment are a standard image as shown in Fig. 4. With the quality as described in Table 1. Noises will be added into the low-quality images that used in the experiments from 1 to 100 dB.

3.2 Super-Resolution Image (SRG)

To improve image quality in terms of resolution and sharpness. Interference filter and the interpolation of pixels are given most accuracy using either color estimation methods or probability of color values for a new pixel by eliminating pixels with different color values within the image area size of 3 × 3 pixels as Eq. (6). Then insert a new pixel with the appropriate color value instead of pixels that are eliminated from the image compare to the original image. The method for pixel removal and color estimation for the pixels show as follow.

$$W_i = \begin{bmatrix} Z_{11} & Z_{12} & Z_{13} \\ Z_{21} & Z_{22} & Z_{23} \\ Z_{31} & Z_{23} & Z_{33} \end{bmatrix} \tag{6}$$

$$c_i = \frac{|Z_i - \bar{Z}_i|}{\bar{Z}_i} \tag{7}$$

$$p_i = \frac{c_i}{\sum_{i=1}^{n} c_i} \quad (8)$$

Where

p_i = Interpolation value for pixel i,
c_i = Value of pixel i
Z_i = Pixel of M × N in W_i,
$\bar{Z}_i$ = Average of pixels

The interpolation pi value from Eq. (8) maybe close to zero if considered in the informational field, it can be the level of uncertainty to find the closest image to have a value approaching zero, Therefore, when measuring the uncertainty of pixels that are close together or cannot be determined, Eq. (9) as follow should be applied.

$$P_i = -\sum_{i=1}^{n} p_i \log p_i \quad (9)$$

Where

P_i = Interpolation values of the new pixel.

The pixels *i* in areas that very different the P_i value is high because the probability of interest in that point is high. The idea is to convert a pixel value into a probability value that represents the difference in a particular area. In the scope of M × N, it will be processed to improve the pixel to get pixels that are close to each other, if the P_i value is low, it will not be processed to improve pixel. The step of developing SRG algorithm for image processing is shown as follow.

SRG Algorithm:1

```
1. input image;
2. set variable;
3. create Matrix zero(M,N,d);
4. loop
        read number into variable
        For x = 0,x<M; For y = 0,y<N;
                ZBar = MeanWindows(x,y);
                ZB(x,y) = [Z(x,y) - ZBar / ZBar];
                p = SumWindows(x,y);
        if (probability p = Low)
                (x,y) = ZB(x,y);
        End
        If (probability p = Hight)
                (x,y) = RemoveNoise;
        else
                (x,y) = Z(x,y);
        End
5. End loop;
```

Algorithm 1, input image, Z (x, y) is the pixel at the position (x, y) of the image and the size is M × N the position of (x, y) is x = 0, 1, 2,…, M − 1, y = 0, 1, 2,…, N − 1. With W_i as the area for limiting to be processed with the matrix size k × k. For this research, given k = 3, therefore, finding the probability of eliminating different values can be obtained from this point. The average value for estimating the value of the position (x, y) using the MeanWindows function. For the sum of averages and finding the probability difference calculated with in W_i position (x, y) according to Eqs. (8) and (9), will use the SumWindows function. When the probability for the different values is obtained, then the position will be eliminated and insert a new pixel that has been estimated instead of being removed by the RemoveNoise function. For image from magnification, reduce blur and noise increase the sharpness of the image. The performance of the Bicubic technique has been improved in some parts. In order to improve image quality with greater resolution and sharpness, Eq. (10) will be used as follow.

$$\widehat{PM_x} = (P_x - \lfloor P_x \rfloor \times NN) \tag{10}$$

$$\widehat{PM_y} = \left(P_y - \frac{\lfloor P_y \rfloor}{NN}\right) \tag{11}$$

Where $\widehat{PM_x}$ and $\widehat{PM_y}$ is the point that you want to find value.

P_x, P_y is the point to find the distance the x, y.

Equations (10) and (11) are improved equation for the processing of the pixel that wanted to interpolation. The new interpolation pixel will be adjusted to the coefficients of each pixel with a new image in the x, y. To add sharpness to the image you want to enlarge image. For the proposed technique, the image processing data from the original image is used and create a new image with a matrix of 4 × 4 with 3 dimensions according to the RGB pattern. P_x, P_y will be re-created based on the image size to resize. Where $\lfloor P_x \rfloor$ is the level of processing in the Row with size 4 row, $\lfloor P_y \rfloor$ is the level of processing in the column with size 4 column, *NN* is the number of neighbors surrounded by a value of 4. Increased pixel estimation to get the most accurate pixel value.

Table 2. Evaluation results for the Lena at 3x.

Noise (dB)	Bicubic		HRI	
	PSNR	SSIM	PSNR	SSIM
10	22.303	0.562	22.696	0.577
20	21.432	0.429	21.890	0.455
30	20.325	0.328	20.850	0.355
40	19.193	0.257	19.767	0.283
50	18.142	0.207	18.744	0.231

The experimental results in terms of PSNR and SSIM shown in Table 2. The contract values are set to increase from 1–50 dB, the proposed can be used to improve image quality or reduce noise better than Bicubic technique.

4 Experimental Results

Enhancement of the proposed with Bicubic, ASDS_AR_NL [22] and Demo SR [23] are evaluated by PSNR and SSIM values of newly created images and low-quality used in the experiment from high-quality original image. The image without sharpness or blur were used in the experiment by setting as standard to reduce the quality of the original image with MATLAB. Blur = fspecial ('gauss', 11, 5) and Noise = (Gaussian noise, 1–100(dB)). This research will not be evaluated in the field of blurring. To improve the image quality from low-resolution images. To create high-resolution images with PSNR and SSIM that can be separated according to the original image used in all experiments according to Tables 3, 4, and 5 respectively. The results of images obtained from the experiments as shown in Figs. 5 and 6. The experiment of creating high-resolution images with the proposed method, when noises are included in image without sharpness or blurred image. The improvement of image quality with this technique shown in Tables 3, 4 and 5. As the results, it found that when the image quality was lowered, the proposed decreased with the quality of the original image but there is not much decrease in value compared to other techniques used in the experiment. The average PSNR and SSIM is higher than other techniques used in the image this experiment. From the experiment, the sample images obtained from the experiment in Figs. 5 and 6 can be viewed with the naked eye, also can see the difference of the image, especially the image that has a lot of noise.

Table 3. Evaluation results for the Lena at 3x.

Noise (dB)	ASDS_AR_NL		Demo_SR		SRG	
	PSNR	SSIM	PSNR	SSIM	PSNR	SSIM
20	20.899	0.428	20.273	0.304	21.890	0.455
40	12.608	0.099	17.212	0.160	19.767	0.283
60	9.543	0.040	15.156	0.106	17.821	0.192
80	8.573	0.022	13.686	0.077	16.316	0.142
100	8.138	0.018	12.696	0.061	15.187	0.112
Average	11.952	0.121	15.804	0.141	**18.196**	**0.236**

Table 4. Evaluation results for the Parrots at 2x.

Noise (dB)	ASDS_AR_NL		Demo_SR		SRG	
	PSNR	SSIM	PSNR	SSIM	PSNR	SSIM
20	20.391	0.385	19.900	0.258	21.510	0.444
40	11.107	0.061	16.346	0.111	19.375	0.237
60	8.869	0.026	14.154	0.068	17.419	0.149
80	8.123	0.019	12.712	0.048	15.902	0.104
100	8.050	0.024	11.751	0.037	14.773	0.080
Average	11.308	0.103	14.972	0.104	**17.795**	**0.202**

Table 5. Evaluation results for the Bike at 2x.

Noise (dB)	ASDS_AR_NL		Demo_SR		SRG	
	PSNR	SSIM	PSNR	SSIM	PSNR	SSIM
20	17.467	0.355	16.886	0.211	17.630	0.292
40	10.577	0.105	14.760	0.122	16.640	0.205
60	8.404	0.046	13.116	0.085	15.491	0.152
80	8.065	0.028	11.952	0.064	14.460	0.119
100	8.487	0.033	11.150	0.051	13.616	0.097
Average	10.600	0.113	13.572	0.106	**15.567**	**0.173**

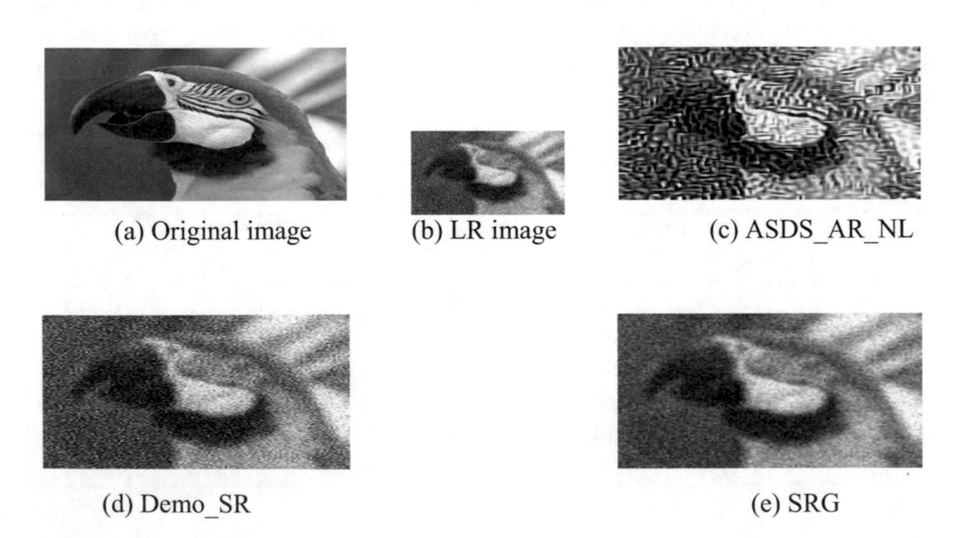

Fig. 5. Examples of Parrots image, LR image noise 30 dB.

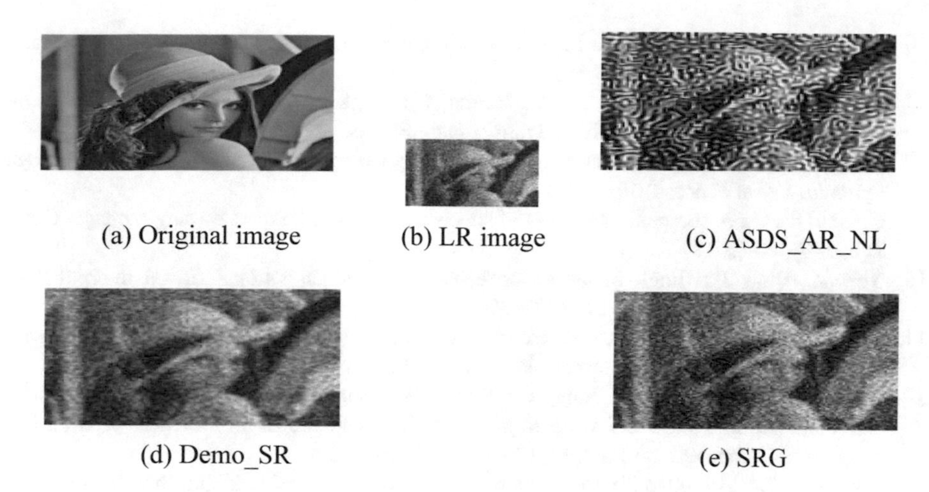

Fig. 6. Examples of Lena image, LR image noise 40 dB.

5 Conclusion

This paper proposed an Image Super-Resolution technique to enhance the quality of image by adjusting coefficient of pixel using the estimation value of interpolation from neighboring pixels. This technique is developed from the interpolation methods of Bicubic technique in order to improve image quality in terms of resolution and sharpness. A new image will be created based on Image Super-Resolution (SRG) to reconstructs a higher-resolution image. SRG technique calculate a value from neighboring pixels to create new image with high resolution and sharpness. For the experimental results, SRG is a good technique that can be used to improve a quality of image from low quality image. Therefore, a low-quality image with a lot of noises and blurred can be improved by replacing new image using SRG technique.

References

1. Ganga, M.P., Prakash, C., Gangashetty, S.V.: Bessel transform for image resizing. In: 2011 18th International Conference on Systems, Signals and Image Processing (IWSSIP), pp. 1–4 (2011)
2. Hsin, H.: Combination of saliency histogram equalization and seam carving for image resizing. J. Eng. **7**, 285–291 (2017)
3. Zhiwei, H., Mingyu, G., Haibin, Y., Xueyi, Y., Li, Z., Guojin, M.: A new improved seam carving content aware image resizing method. In: Industrial Electronics and Applications, pp. 738–741 (2013)
4. Lehmann, T.M., Gonner, C., Spitzer, K.: Survey: interpolation methods in medical image processing. IEEE Trans. Med. Imaging **18**, 1049–1075 (1999)
5. Yue, L., Shen, H.: Image super-resolution: the techniques, applications, and future. Signal Process. **128**, 389–408 (2016)

6. Nasrollahi, K., Moeslund, T.B.: Super-resolution: a comprehensive survey. Mach. Vis. Appl. **25**, 1423–1468 (2014)
7. Zhang, Y., Fan, Q., Bao, F., Liu, Y., Zhang, C.: Single-image super-resolution based on rational fractal interpolation. IEEE Trans. Image Process. **27**(8), 3782–3797 (2018)
8. Hu, J., Wu, X.: Noise robust single image super-resolution using a multiscale image pyramid. Signal Process. **48**, 157–171 (2018)
9. Ga, P., Um, G.: A compendious study of super-resolution techniques by single image. Optik **166**, 147–160 (2018)
10. Yun, Z., Qing, F.: Single-image super-resolution based on rational fractal interpolation. IEEE Trans. Image Process. **27**, 3782–3797 (2018)
11. Xiao, J., Wang, Y.: Adaptive shock filter for image super-resolution and enhancement. J. Vis. Commun. Image Represent. **40**, 168–177 (2016)
12. Li, H., Gao, Y., Dong, J., Feng, G.: Super-resolution based on noise resistance deep convolutional network. In: Proceedings of the 6th International Conference on Bioinformatics and Computational Biology, Chengdu, China (2018)
13. Dengwen, Z.: An edge-directed bicubic interpolation algorithm. In: 3rd International Congress on Image and Signal Processing, Yantai, pp. 1186–1189 (2010)
14. Patanavijit, V.: Super-resolution reconstruction and its future research direction. AU J. Technol. **12**(3), 149–163 (2009)
15. Zhao, X., Yang, R., Qin, Z., Wu, J.: Study on super-resolution reconstruction algorithm based on sparse representation and dictionary learning for remote sensing image. In: 10th International Congress on Image and Signal Processing, BioMedical Engineering and Informatics (CISP-BMEI), Shanghai, pp. 1–4 (2017)
16. Yang, X., Wu, W., Liu, K., Kim, P.W., Sangaiah, A.K., Jeon, G.: Long-distance object recognition with image super resolution: a comparative study. IEEE Access **6**, 13429–13438 (2018)
17. Agustsson, E., Timofte, R.: NTIRE 2017 challenge on single image super-resolution: dataset and study, pp. 1122–1131 (2017)
18. Timofte, R., Agustsson, E., Gool, L.V., Yang, M.-H., Zhang, L., Lim, B.: NTIRE challenge on single image super-resolution: methods and results, pp. 1110–1121 (2017)
19. Yun, C., Yan, W., Qi, L.: Research on digital image scaling based on bicubic filter algorithm. In: 2018 IEEE 3rd International Conference on Image, Vision and Computing (ICIVC), Chongqing, pp. 225–229 (2018)
20. Motmaen, M., Mohrekesh, M., Akbari, N., Karimi, N., Samavi, S.: Image inpainting by hyperbolic selection of pixels for two-dimensional bicubic interpolations. In: Iranian Conference on Electrical Engineering (ICEE), Mashhad, pp. 665–669 (2018)
21. Jaya, V.L., Gopikakumari, R.: IEM: a new image enhancement metric for contrast and sharpness measurements. Int. J. Comput. Appl. **79**(9), 1–9 (2013)
22. Dong, W., Zhang, L., Shi, G., Wu, X.: Image deblurring and super-resolution by adaptive sparse domain selection and adaptive regularization. IEEE Trans. Image Process. **20**(7), 1838–7857 (2011)
23. Wang, Z., Liu, D., Yang, J., Han, W., Huang, T.: Deep networks for image super-resolution with sparse prior. In: Deep Networks for Image Super-Resolution with Sparse Prior, pp. 370–378 (2015)

Smart Telematics System with Beacon and Global Positioning System Technology

Krutpong Krunthep, Siranee Nuchitprasitchai(✉), and Yuenyong Nilsiam

King Mongkut's University of Technology North Bangkok, Bangkok, Thailand
krutpong@gmail.com, siranee.n@it.kmutnb.ac.th, yuenyong.n@eng.ac.th

Abstract. Road accident is the first cause of death for people ages 15–29 around the world. In Thailand, the statistic shows that the road accident increases every year as well as the number of injured and death due to the road accident. This research proposes a system consists of mobile application, beacon, and GPS system to record the information. When application detects the beacon in the car, the status of driver is recorded as active in a car. The route information is tracked using GPS system of the smartphone. All information is sent to the cloud database via the internet connection of the smartphone and the information can be viewed on the smartphone application. The application was implemented successfully and working as expected. The system can detect the status of the drivers and the route sending to the cloud for further analyzing. Based on the information, many insightful information can be found, and many useful predictions can be made.

Keywords: Telematics · Beacon · GPS · Global Positioning System

1 Introduction

Road accident is one of the major problems in the world which cause loss of life, injury, and handicapped. According to the global status report on road safety 2015 [1], it is estimated that 1.25 million people of world population die in road accident each year or more than 3,400 deaths each day. Moreover, the number of people who are injured and handicapped is about 20–50 million accumulatively. Finally, it is the first cause of death for the world population ages 15–29 and the number of males is three times of females.

In Thailand, the statistic of logistics and transportation safety [2] indicates that most of road accidents happen to the three types of the personal vehicles, motorcycle, car, and pickup truck. They are in the trend of increasing every year from 2010 to 2017. Especially in the last five years, accidents of all three types rise up more than 50% (see Fig. 1).

P. Meesad and S. Sodsee (Eds.): IC^{2}IT 2020, AISC 1149, pp. 181–189, 2020.
https://doi.org/10.1007/978-3-030-44044-2_18

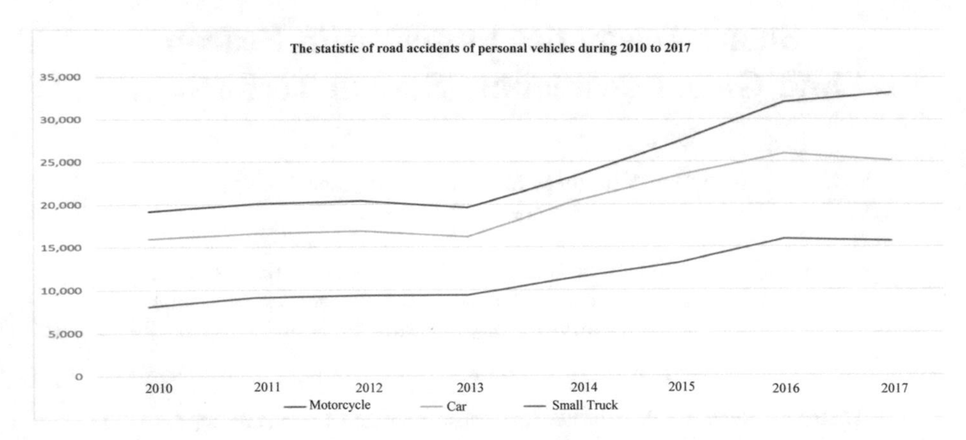

Fig. 1. The statistic of road accidents of personal vehicles during 2010 to 2017.

Moreover, according to the same source [2], the number of people who are injured is more than 20,000 per year and the number of deaths is 4,000 to 5,000 per year. Approximately, the value of loss from road accidents is more than 200 million baht per year. In Fig. 2, the statistic shows that the number of deaths in men is 30% more than in women from road accidents.

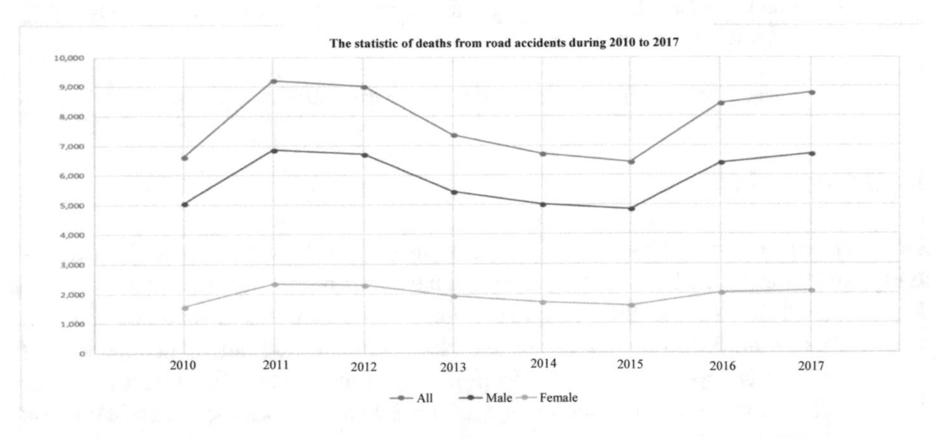

Fig. 2. The statistic of deaths from road accidents during 2010 to 2017.

At present, the technology has been developed exponentially. One of the recent technologies is a smartphone which becomes more and more as a basic need in life. The usage of smartphone has been increased dramatically each year [3]. In 2018, about 70% of population of Thailand use smartphones [4].

Smartphone has much more features compare to the earlier mobile phone. More than just making a call, it can connect to the Internet, connect to other devices for sending messages, and connect to a speaker via Bluetooth. Another device that the smartphone can connect to is beacon via Bluetooth Low Energy (BLE) to receive the signal from the beacon and estimate the distance based on the strength of the signal. A smartphone also has a Global Positioning System (GPS) receiver which communicates with satellites to identify the position on the earth. These are the basic features of smartphones.

From an article, iBeacon is a promising technology for positioning system based on signal strength and supported by iOS and Android [5]. According to the real use of iBeacon and Android mobile application to identify the location in Khon Kaen University (KKU) Virtual Museum and display images and videos and also audios according to the location in the museum [6]. The satisfaction is good for the accuracy of the location. This is one of examples that using iBeacon and mobile application together for specific purpose. It should be the proof that the technology is mature enough for real use.

From the problem of road accident and the technology in the previous paragraph, a Smart Telematics system with Beacon and Global Positioning System Technology is proposed for recording when driver getting in and off a car and the coordinates of the car using GPS which will be sent to the cloud server for road accident analysis. If there is a high probability of road accident, the client will be contacted to offer helps immediately. For example, an accident happens in the night and no one would know until the morning. If the system can notify a staff to check on the client and reports for emergency helps in case of needed. This would help reduce the loss of lives significantly.

2 Literature Review

iBeacon is a beacon protocol developed by Apple and released in Apple Worldwide Developers event. It gains popularity in the use for retail, museum, and research. Indoor positioning system using iBeacon for advertisement of Faculty of Information Technology [7] was set up by installing beacon devices as in museum in some countries. The result shows that the accuracy of location using Beacon is more than 90%. Beacon technology is supported by iOS and Android with Bluetooth Low Energy or Bluetooth 4.0. The two operating systems are the main operating system for mobile devices.

From a research about broadcasting protocol involving trusting positioning in wireless ad-hoc communication network for vehicles [8], beacon is working by automatically broadcasting information in frequency of 2.4 GHz to all receiving devices. Another study of adjustable beacon for wireless ad-hoc communication network [9], testing by all cars finding the signal of beacon from other cars. The result shows that the stability of the communication is good in the radius of 10 to 30 m. The information sending includes Universal Unique Identifier (UUID), Major, Miner, Mac Address, Accuracy, and Received Signal Strength Indicator (RSSI). From a research, a desynchronization algorithm for beacon collision avoidance on vehicular networks [10], the collision of beacon was tested with more than three beacon devices and the result shows that the beacon signals cause trouble for each other. In order to solve the problem, a mathematic model needs to be used to find the strongest signal. Another way to solve the problem is to use a mac address which is same as automatically finding a car using iBeacon [11]. The later technique is used by the researchers in this research.

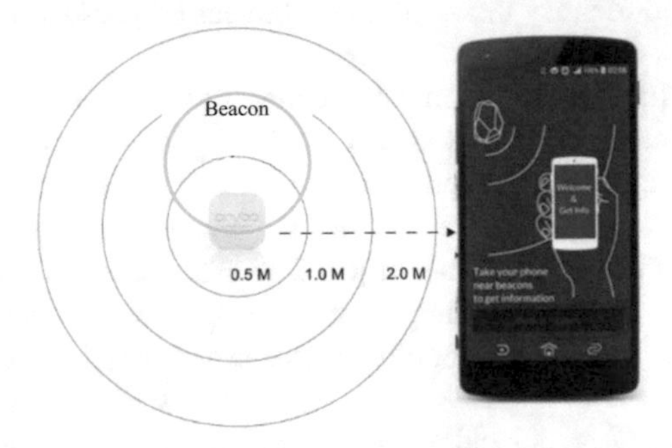

Fig. 3. The characteristics of communication between beacon and smartphone.

From Fig. 3, it is how beacon works. A beacon will keep sending the signal all the time and a smartphone will monitor the signal and receive information from beacons in the area without specifying any device.

Global Positioning System (GPS) is a system to specify a position on the earth using satellite networks of 3 parties including America, Russia, and Europe. NAV-STAR (Navigation Satellite Timing and Ranging GPS) is belong to America and has 28 satellites. 24 of satellites are active and 4 of satellites are ready to active when needed. They are managed by the Department of Defenses. The orbital radius from the earth is 20,162.81 km or 12,600 miles. Each satellite orbits around the earth in 12 h. Galileo is a network of 27 satellites belongs to and is managed by European Satellite Agency (ESA). Global Navigation Satellite or GLONASS is managed by Russia Military Space Force (Russia VKS). Currently, people around the world can use the data from the satellites of America (NAVSTAR) for free according to Access to Information Policy (AIP) of the government of the United States. Therefore, general people can access to the information at the accuracy level without a threat to national security, the error of accuracy within 10 m. Satellite Management includes the main earth station at Falcon Air Force Base in the United States and 5 earth stations are distributed in different continents around the world. A user needs a receiver to receive the signal from the satellites and interprets the data for further processing according to usages. A device receives the signal from the satellite network, such as NAVSTAR which 24 satellites distributing into 6 orbits and 55° from the Equator and from each other, so there are 4 satellites in each orbit. The signal consists of data for position and time of sending the signal. The distance between the satellite and the receiver is calculated by the different of sending and receiving time. At least 4 satellites are needed to calculate the position of the receiver accurately. The distances of 3 satellites are used to calculate the position on the panel and then the distance of the fourth satellite is used to calculate the altitude according to the sphere shape of the earth.

The greater distance between selected satellites for calculating position of the receiver and more satellites selected will result in better accuracy. Many things can affect the accuracy of the position calculation, especially, the variation of the atmosphere including atmospheric electricity, humidity, temperature, and atmospheric pressure. The refraction of the signal on objects makes the signal weaker. The environment around the receiver also affects the position accuracy, such as, cover by a mirror, mist, or leaves due to the refraction. Last but not least, the efficiency of the receiver and its processing power. According to some researches, controlling and tracking vehicles using mobile GPS based on GPRS/3G networking [12] and location-based service on GPS mobile tracking [13], the air is a significant obstacle for a smartphone to receive satellite signal for GPS system, not a speed of the vehicle.

The calculation of the distance between the satellite and the receiver is shown in Eq. 1. The times are from the clock of satellites which are high accuracy down to nanosecond level and they are cross checked with earth stations all the time.

$$distance = speed\ of\ signal \times (receiving\ time - sending\ time) \quad (1)$$

The last component is the actual position of each satellite while sending the signal. The orbits of satellites are defined before sending the satellites into the space and earth stations always monitor the orbits of them for the correction.

According to a research, using open source GPS tracking system with smartphone for truck tracking case study of Tip Khelang drinking water [14], shows that using position data for measuring speed of a vehicle from GPS system of a smartphone running Android operating system by sending the data to a database in every 7, 15, 30 s is best at the rate of 30 s for the efficiency of the system.

Another research, Accuracy Indoor Proximity Zone Detection Based on Time Window and Frequency with Bluetooth Low Energy [15], tests an accuracy of distance calculation based on the strength of the signal from a beacon device. The result shows that the accuracy is in high variation for the distance less than 0.25 m. Therefore, the distance of more than 0.25 m will be used to detect the status of the user.

A research about an acceptance of user to location-based service, The Study of Impact of Technology Acceptance and Perceived Risk on User's Trust Perception of Location-based Services in Bangkok [16], indicates that users consent to provide position data if they understand the benefits and trust the security of the data transferring processes.

3 Methodology

From Fig. 4, Smart Telematics with Beacon and Global Positioning System Technology uses a Bluetooth connection of a beacon installed in a car with a smartphone installed a developed application to identify the status of the user, whether in a car or not. The distance between the beacon and the smartphone needs to be less than 0.5 m and data from the beacon will be sent to a cloud via the internet connection of the smartphone along with position data from GPS system of the smartphone. When the application on the smartphone detects the beacon in the car and the current position data is changing, the data will be sent to the cloud every 30 s as mentioned in the previous section.

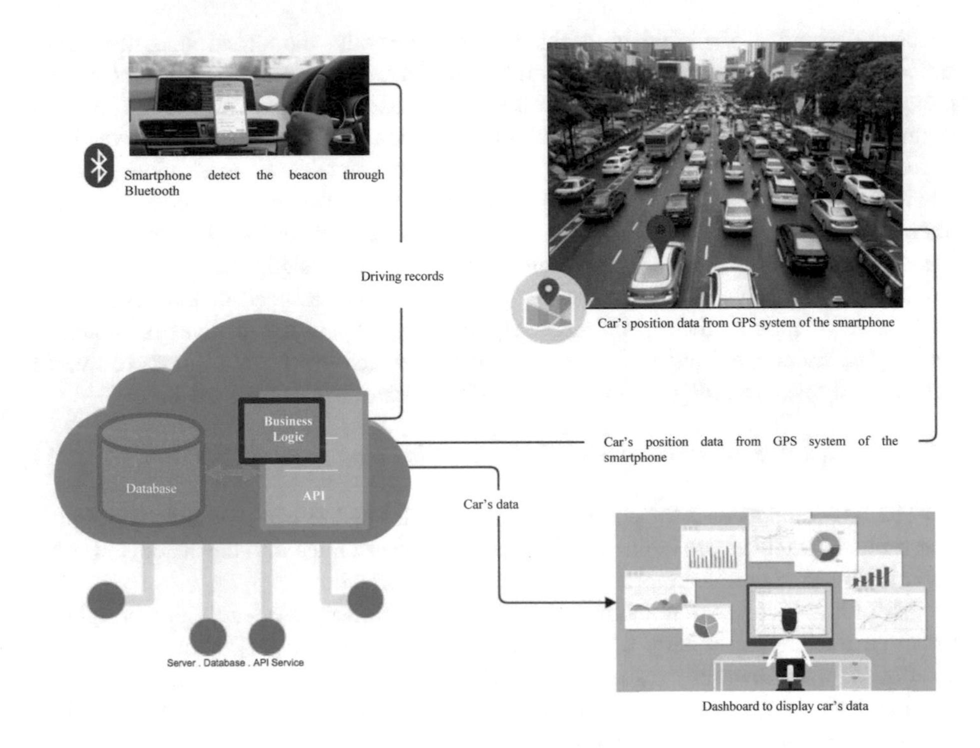

Fig. 4. Overview of Smart Telematics system with Beacon and Global Positioning System Technology.

The target group of users should be interviewed for requirements of the application. The database structure needs to be designed to support all the data. A few Mobile application frameworks should be reviewed and one appropriated should be selected. The architecture of the mobile application should be designed and then implemented. The connection between the beacon device and a smartphone via Bluetooth should be tested as well. The cloud system should be prepared for database management system.

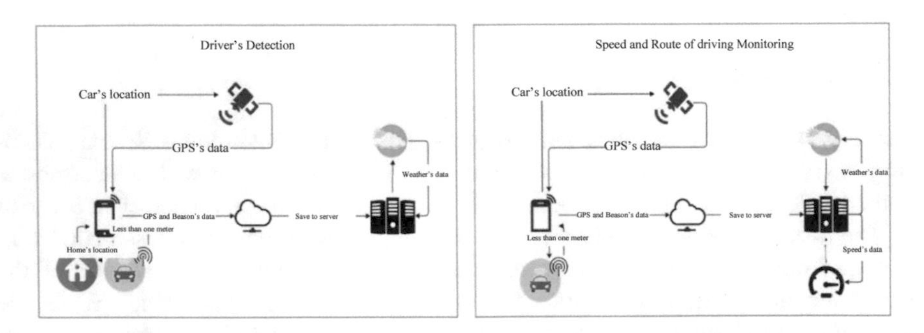

Fig. 5. The operation of detecting the user in a car and the operation of monitoring the speed and the route of driving.

In Fig. 5, the operation of detecting a status of a user and the operation of monitoring the speed and the route of driving are shown in detail. The current speed of a vehicle is processed by the position data of the vehicle in every 30 s.

The security of the system and the privacy of users should be protected by encrypted transferred data between mobile application and the cloud.

Moreover, the purposed system should include the capability to find nearby auto shops from the current location of the user. However, in order to use the system, users need to register and activate their accounts.

4 Results

A beacon device (see Fig. 6) was installed in a car. It can be installed anywhere in the car, so it does not affect the driving.

Fig. 6. A beacon device.

After interviewing users, the requirements were collected as inputs for the design of the application and its database structure. Flutter framework [17] was selected for the application development tool. The connection of the beacon device and the smartphone is working seamlessly. The application on the smartphone consists of member information, driving history, weather information, average speed, and a section for administrator.

User can register to the system via the application and after signed in the user can edit personal information (see Fig. 7). First, a beacon device needs to be register through the mobile application using its identification number. The status of the user is detected by the distance between the beacon and the smartphone. The location data is sent to the cloud via the mobile internet. The data is stored on the cloud for displaying and analyzing.

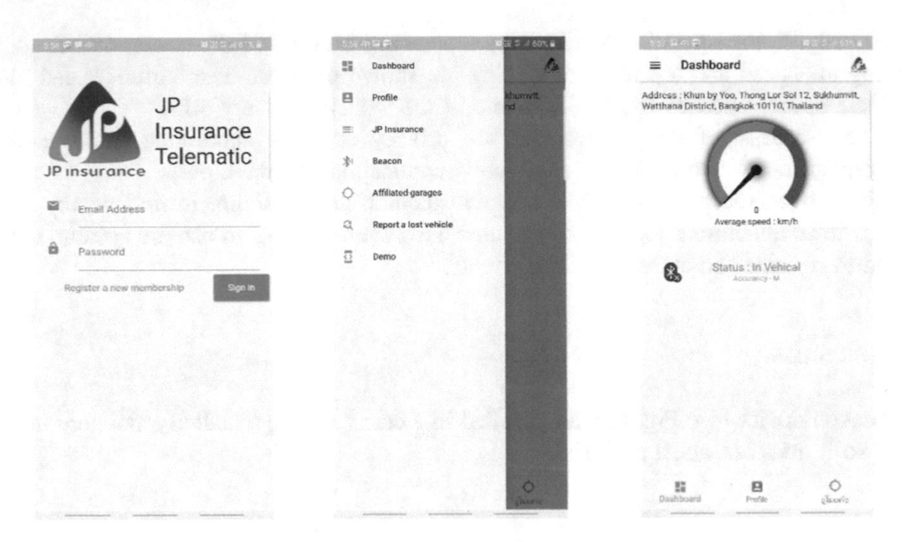

Fig. 7. Some screenshots of the application.

5 Conclusion

The smart telematics system with beacon technology and global position system was successfully developed and implemented. The status of user can be detected using the beacon and the smartphone application and the data of location of the vehicle can be retrieved from the GPS system of the smartphone. Data are sent to the cloud via the mobile internet. The functions of the application are working as expected. User can register the beacon device, edit personal information, look up insurance information, and find a nearby auto shops.

The data stored on the cloud are status of user whether in or off the car, location and surround weather of the user in every 30 s, and insurance information. These data are great information and can be summarized and show to administrators. For the future work, analyzing the data can bring out many insightful information about users and their driving behaviors. Moreover, many predictions can be done based on these data.

Acknowledgement. This study was supported by JP Insurance Public Company Limited.

References

1. Global Status Report on Road Safety 2015. http://www.who.int/violence_injury_prevention/road_safety_status/2015/GSRRS2015_Summary_EN_final.pdf. Accessed 19 Apr 2019
2. The Situation of Traffic Accidents by Type of Road User, Property Damaged, Alleged Offenders, Region: 2011–2018. http://statbbi.nso.go.th/staticreport/page/sector/en/15.aspx. Accessed 19 Apr 2019
3. Number of smartphone users in Thailand from 2013 to 2022. https://www.statista.com/statistics/467191/forecast-of-smartphone-users-in-thailand/. Accessed 01 May 2019

4. Percentage of population 6 years of age and over using Mobile telephone and Smart phone, Computer and Internet by region. http://stiic.sti.or.th/stat/ind-it/it-t012/. Accessed 1 May 2019
5. Introduction to iBeacon the coming technology for future positioning system. https://www.blognone.com/node/57349. Accessed 1 May 2019
6. Bhumi, W.: PULINET J. **4**(3), 249–258 (2017)
7. Meekrin, M.: An indoor location identification system using iBeacon for public relations of faculty of information technology, Information Technology, King Mongkut's University of Technology North Bangkok (2017)
8. Pramuanyat, N.: A location-aware reliable broadcasting protocol in vanet, Department of Computer Engineering, Faculty of Engineering, Chulalongkorn University (2017)
9. Thaina, C.: A study of adaptive beacon transmission on vehicular ad-hoc networks, Department of Computer Engineering, Faculty of Engineering, Chulalongkorn University (2011)
10. Settawatcharawanit, T.: A desynchronization algorithm for beacon collision avoidance on vehicular networks, Department of Computer Engineering, Faculty of Engineering, Chulalongkorn University (2012)
11. Yoksiri, P., Boonrawd, P.: An automated vehicle location prototyping using iBeacon technology. In: The 12th National Conference on Computing and Information Technology (NCCIT 2016), pp. 521–526. King Mongkut's University of Technology North Bangkok, Faculty of Information Technology, Bangkok (2016)
12. Kaewkiriya, T.: Controlling and tracking vehicles using mobile GPS based on GPRS/3G networking. Romphruekj **32**(2), 85–102 (2014)
13. Romsaiyud, W.: Location-based services on GPS mobile tracking, Graduate School of Information Technology, Siam University (2012)
14. Sawangtook, W.: Using open source GPS tracking system with smartphone for truck tracking case study of Tip Khelang drinking water. Ind. Technol. Lampang Rajabhat Univ. J. **7**(1), 40–51 (2013)
15. Moungsiri, N., Chadarattanatitie, P.: Restaurant search phase Android device with GPS system. In: The 2nd National Conference on Technology and Innovation Management (NCTIM 2016), pp. 463–470. Rajabhat Maha Sarakham University, Maha Sarakham (2016)
16. Rungruangsak, J.: The study of impact of technology acceptance and perceived risk on user's trust perception of location-based services in Bangkok. Bangkok University (2015)
17. Flutter. https://flutter.dev. Accessed 19 June 2019/

Incremental Object Detection Using Ensemble Modeling and Deep Transfer Learning

Piyapong Huayhongthong, Siriyakorn Rerk-u-suk(✉), Songwit Booddee, Praisan Padungweang, and Kittipong Warasup

School of Information Technology, King Mongkut's University of Technology Thonburi, Bangkok, Thailand
{piyapong.bestiiz, siriyakorn.sryk, songwit.mik}@mail.kmutt.ac.th, praisan@gmail.com, kittipong@sit.kmutt.ac.th

Abstract. Object detection is a subset of computer vision that can be accomplished using machine learning. The main process of object detection using machine learning model is model training with images containing objects of interest. However, the model training need a lot of training images. In addition, to improve the model ability to detect addition class of object, it need to be re-trained with both old and new image datasets. It is a time and computation consuming process. This paper proposes an incremental object detection model without re-training the old images. An ensemble model and transfer learning approach are used. The proposed model consist of three parts, two object detection sub-models and a decision model, which are a pre-trained model, a transferred-model and an ensemble model respectively. To illustrate the proposed model, the trained YOLO algorithm training with eighty object categories, 330,000 total images, from COCO image dataset is selected as the pre-trained model. It also be used as an initial model to train the transferred-model using transfer learning technique. Only new images are used for transferred-model training. The ensemble model with the bagging technique is used as a final classifier for choosing the best decision from both sub-models. Using our proposed model, the need of both the number of training dataset and the training time are reduced. Only several hours are needed for model training with three new object categories, 3,000 total images. The experimental results show that the proposed model achieve high performance on test image dataset with 93.33% accuracy.

Keywords: Common object in object (COCO) · Deep transfer learning · Ensemble modeling · Object detection · You Only Look Once (YOLO)

1 Introduction

Machine learning is used in many aspects in our daily life, e.g., artificial intelligence, speech recognition, data analysis, object detection, etc. It is a technology that allows computers to learn for solving problem of interest by feeding data and target called training data. The performance of machine learning depends on the training data, model

P. Meesad and S. Sodsee (Eds.): IC^{2}IT 2020, AISC 1149, pp. 190–198, 2020.
https://doi.org/10.1007/978-3-030-44044-2_19

architecture, and learning algorithm. It successfully be applied in object detection [1, 2], a subset of computer vision.

Object detection is extensively used in many applications such as objects tracking, anomaly detection, people counting, face detection [2], etc. Many researches use deep learning technique in object detection applications [3–6]. The main process of object detection using the machine learning model is model training with images containing object of interest called dataset. Each image in the dataset may contain one or more objects. Many images are required for training a deep learning model. Consequently, the training processes consume a lot of computation and time. After a proper training approach, the model may ready to use for detecting objects of interest which belong to the class presenting in the training set. However, if we need to train model to detect other classed of objects which are not presented in the training dataset, we need to re-train the model by including the images containing objects in the new classes. The new training dataset contains both previous training images and new images. Obviously, it is a time and computation consuming process for re-training to build an incremental object detection model.

This paper applies an ensemble method to build an incremental object detection model without re-training any well-trained model. The new dataset is needed only for training a sun-model. The proposed model consist of three parts including a well pre-trained model, a transferred model for detecting additional objects, and an ensemble model for determining final decision. The transferred model simply serves as the incremental object detection part by using transferred learning approach. This approach reduces the time and the required number of images for model training. The ensemble model is implemented using bootstrap aggregating (Bagging) technique [7].

This paper is organized as follows: Sect. 2 reviews object detection using deep learning. The proposed model is presented in Sect. 3 explaining how each sub-models work. Then the experimental results of the proposed model are shown in Sect. 4. Finally, Sect. 5 is the conclusion.

2 Object Detection using Deep Learning

The main role of object detection is to detect objects in the images and to recognize the class of those objects. The output may simply locates each detected objects in the image with bounding boxes, rectangular box around the objects together with the name of the object. The input of the object detection model can be an image or video containing one or more objects.

There are many popular object detection models such as the R-CNN family (R-CNN [8], Fast R-CNN [9], Faster R-CNN [10], and Mask R-CNN [11]) and You Only Look Once (YOLO) [12]. These models are different in terms of architecture, training algorithm and number of parameter.

Regions with Convolutional Neural Network Features (R-CNN) [2, 8] is the basic technique for object detection but time consuming. The input of R-CNN is an image which is extracted into small sizes called region proposals. R-CNN uses a selective search method to extract the region proposals, regions are possibly the region of object or part of object. The selective search provides a set of candidate region proposals.

Then all these region proposals are wrapped and sent to the convolutional neural network (CNN) [13], a feature extractor, one by one for extracting features. Then Support Vector Machine (SVM) is used to classify the presented object.

The R-CNN need to feed all region proposals into CNN one by one which is a time-consuming process. Fast R-CNN [2, 9] feeds the whole image into CNN to generate a convolutional feature map. It uses region of interest pooling layer to generate a fixed size vector of region proposals. Therefore, it can be presented to a fully connected (FC) layer. The Softmax function [14] is used in the FC layer to classify the objects.

Faster R-CNN [2, 10] is unchanged from Fast R-CNN much, the input image is fed into CNN to generate a feature map. The region proposal network (RPN) is used for generating region proposals instead of a selective search method in R-CNN and Fast R-CNN. RPN reduces the time-consuming in generating region proposals process. Then region proposals are presented to the pooling layer to generate a fixed size feature.

Mask R-CNN [2, 11] use the same basic structure as Faster R-CNN. In Mask R-CNN the pooling layer is changed into the RoIAlign layer. RoIAlign layer makes Mask R-CNN able to align pixel-to-pixel and has a better instance segmentation while the pooling layer in Faster R-CNN unable. Therefore, Mask R-CNN not only predicts and presents the location of the object with bounding boxes but also generates the segmentation mask on each object in the images.

You Only Look Once (YOLO) [2, 12] is one of the well-known object detection algorithms. It is faster, stronger, and higher performance than all the previous R-CNN family. R-CNN family need to generate many region proposals and the feed them into CNN for predicting and locating the objects. YOLO does not need to generate the region proposals, it uses a single convolutional network and "only looks once" algorithm at the image. The image is divided into grid cells then bounding boxes are generated. The model computes a class probability for identifying object in each bounding box.

Object detection using deep learning is wildly used and the performance is incrementally be improved. However, if we need those models detect new classes of objects, we need to re-train the models to recognize both old classes and new classes. Usually, training a deep learning model requires a lot of training data due to the huge number of model parameters. The object detection using deep learning has no exception. It is a time and computation consuming process. Therefore, this research aims to propose incremental object detection model by avoiding model re-training.

3 Proposed Model

In this paper, we proposed model for improving the process to incrementally detect new objects using transfer learning and a bagging technique. Our proposed model consists of two object detection sub-models and a decision model which are a pre-trained model, a transferred-model and an ensemble model as shown in Fig. 1.

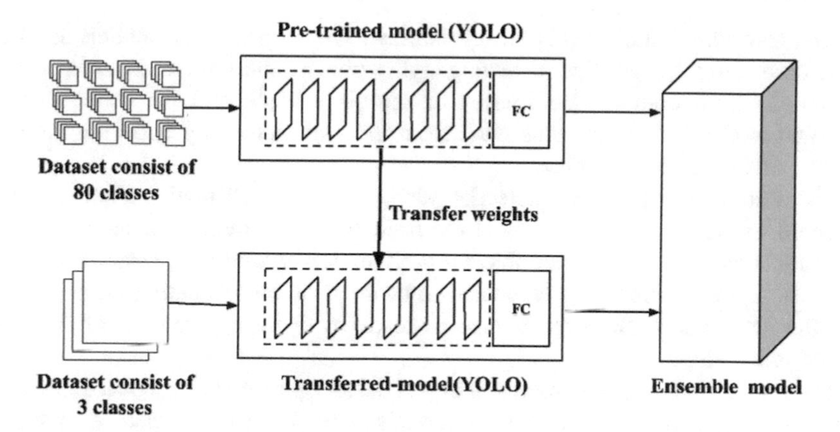

Fig. 1. Proposed model for incremental object detection with YOLO as a pre-trained model.

3.1 Pre-trained Model

The first part of this model is the pre-trained model as shown at the first row in the Fig. 1. Many pre-trained models were trained by many algorithms, for example, R-CNN [8], Fast R-CNN [9], Faster R-CNN [10], etc. All models discussed in Sect. 2 can be used, however, the You Only Look Once (YOLO) [12] algorithm is selected in the experiment. The old training dataset has eighty object categories, 330,000 total images, the COCO image dataset [15]. The output of the model are probabilities of having specific objects in image.

3.2 Transferred-Model

The second part of this model is the transferred-model as shown at the second row in the Fig. 1. This sub-model is created by training only additional dataset containing new classes, which are not presented in the pre-trained model. The transfer learning approach is used here. The first part of pre-trained model serves as feature extractor extracting features of interest from images. The feature extractor of a well pre-trained model can be used for extracting any image and transferred to use in our model. The second part, the Fully Connected layer: FC, serves as a classifier. The classifier need to re-create and re-train for classifying new classes. The additional dataset are used in the transferred-model consisting of three object classes, e.g., camera, frog, and whiteboard. There are around a thousand images of the object per class. The datasets are taken from the Open Images V5 [16].

3.3 Ensemble Model

The last part of the proposed model is the ensemble model as shown at the right most model in the Fig. 1. It combines the results from the two models for making the decision. The ensemble model uses bootstrap aggregating (Bagging) [7] technique. Bagging technique is widely used for model training using a copy of single base learning algorithm but different training datasets. The proposed sub-models use YOLO as the base learning algorithm, detecting eighty objects and three objects respectively. The results from both models consist of the class name and the probability. Our ensemble model decides the final class from the two models by choosing the probability results with highest value.

There are two steps object detection using the proposed model. Firstly, image is presented to the two sub-models and the probability of having objects are detected. There are at most two results per object if both models recognize the same object in the image. Next, the ensemble model decides final decision by comparing the probabilities from the two models. The decision can be made by choosing the class of object with highest probability.

The Fig. 2 illustrates the process of object detection using our proposed model. The frog image is presented to the two sub-models. The object is detected and the probability of being the detected class is computed. The pre-trained model detects object as rat with probability of 0.5695. In this example, the transferred-model detects object as frog with probability of 0.9978. The results are then present to the ensemble model for the final decision. Obviously, the ensemble model should select the highest probability class as the final decision.

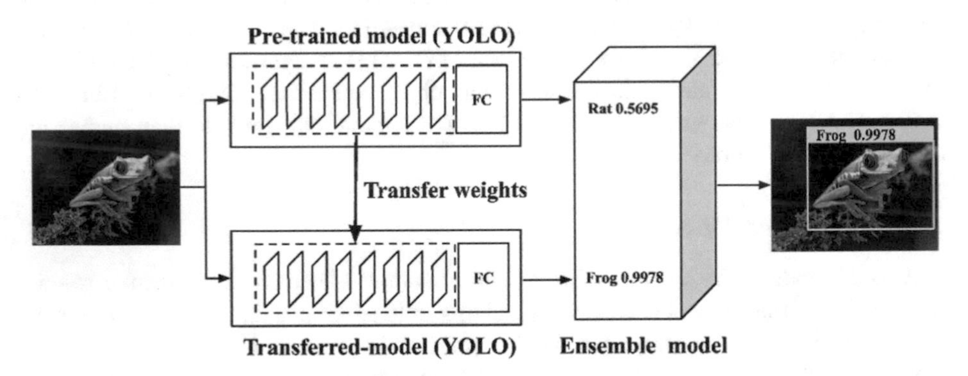

Fig. 2. Example of the prediction processes of the proposed model process.

4 Results

After we trained our sub-model with new datasets, we measure performance of object detection using the unseen test dataset. It is a dataset containing images that are not presented in the training set. We use the same test dataset for testing all models. The test set image contains objects belong to one of six classes including the three classes from the pre-trained model; carrot, knife, and person and three classes from the transferred-model; camera, frog, and whiteboard. There are 120 total images, twenty images for each class. The results are divided into three parts including the results of object detection from the pre-trained model, the transferred-model and the ensemble model.

For the pre-trained model, we test it with the test dataset contain only three objects from the pre-trained model (carrot, knife, and person) as shown in Table 1. The pre-trained model can detect and predict 52 images correctly, predict two images wrongly, and cannot detect six images. One of the wrong prediction objects are knife, the pre-trained model detects and predicts as a remote instead. The total collect prediction is 52 objects, therefore, the accuracy is 86.6%.

Table 1. The result of pre-trained model with test dataset containing three classes.

Predicted	Actual			Total
	Person	Knife	Carrot	
Person	**18**	0	0	18
Knife	0	**15**	0	15
Carrot	0	0	**19**	19
Cup	0	0	1	1
Remote	0	1	0	1
Cannot detect	2	4	0	6
Total	20	20	20	60

Then, we test the pre-trained model with the test dataset containing six objects as shown in Table 2. The pre-trained model cannot predict a camera, frog, and whiteboard correctly. It predicts those as other with low probability. This due to the model can recognize only the objects that it is trained for. The accuracy of this experiment is 43.3%.

For the transferred-model, the test process is the same as the pre-trained model. The transferred-model is tested with the dataset containing the images of a camera, frog, and whiteboard. The transferred-model can detect and predict all images correctly so the accuracy is 100% as shown in Table 3.

We also test the transferred-model with the test dataset containing six objects. The results are shown in Table 4. The transferred-model can detect and predict all trained objects but cannot detect carrot, knife, and person or predict wrongly. For example, an actual image is a person but the transferred-model predict it as a frog. Therefore, the accuracy of this experiment is 50%.

Finally, the ensemble model is tested with the test dataset contains six objects including camera, carrot, frog, knife, person, and whiteboard. Each class has twenty images. The results are shown in Table 5. The ensemble model can detect and predict 112 images correctly, predict three images containing knife and carrot wrongly and cannot detect five images containing person and knife. The accuracy of object detection from this model is as high as 93.3%.

Table 2. The result of pre-trained model with test dataset containing six classes.

Predicted	Actual						Total
	Person	Knife	Carrot	Frog	Whiteboard	Camera	
Person	**18**	0	0	2	5	4	29
Knife	0	**15**	0	0	0	0	15
Carrot	0	0	**19**	0	0	0	19
Frog	0	0	0	0	0	0	0
Whiteboard	0	0	0	0	0	0	0
Camera	0	0	0	0	0	0	0
Remote	0	1	0	0	0	0	1
Cup	0	0	1	0	0	0	1
Elephant	0	0	0	1	0	0	1
Cow	0	0	0	1	0	0	1
Baseball	0	0	0	1	0	0	1
Chair	0	0	0	0	1	0	1
Bottle	0	0	0	0	1	1	2
Laptop	0	0	0	0	3	1	4
Refrigerator	0	0	0	0	1	0	1
Oven	0	0	0	0	0	1	1
Clock	0	0	0	0	0	1	1
TV monitor	0	0	0	0	0	1	1
Cannot detect	2	4	0	15	9	11	41
Total	20	20	20	20	20	20	120

Table 3. The result of transferred-model with test dataset containing three classes.

Predicted	Actual			Total
	Frog	Whiteboard	Camera	
Frog	**20**	0	0	20
Whiteboard	0	**20**	0	20
Camera	0	0	**20**	20
Cannot detect	0	0	0	0
Total	20	20	20	60

Table 4. The result of transferred-model with test dataset containing six classes.

Predicted	Actual						Total
	Person	Knife	Carrot	Frog	Whiteboard	Camera	
Person	0	0	0	0	0	0	0
Knife	0	0	0	0	0	0	0
Carrot	0	0	0	0	0	0	0
Frog	1	2	11	**20**	0	0	34
Whiteboard	0	0	0	0	**20**	0	20
Camera	6	7	0	0	0	**20**	33
Cannot detect	13	11	9	0	0	0	33
Total	20	20	20	20	20	20	120

Table 5. The result of ensemble model with test dataset containing six classes.

Predicted	Actual						Total
	Person	Knife	Carrot	Frog	Whiteboard	Camera	
Person	**18**	0	0	0	0	0	18
Knife	0	**15**	0	0	0	0	15
Carrot	0	0	**19**	0	0	0	19
Frog	0	1	0	**20**	0	0	21
Whiteboard	0	0	0	0	**20**	0	20
Camera	0	0	0	0	0	**20**	20
Remote	0	1	0	0	0	0	1
Cup	0	0	1	0	0	0	1
Cannot detect	2	3	0	0	0	0	6
Total	20	20	20	20	20	20	120

5 Conclusion

In this paper, we proposed an ensemble model that aims to incrementally detect new classes, reduce the time and computation for model training. It can detect more new classes without retraining well-trained model. The proposed model consist of two sub models and an ensemble model. We select YOLO as a base model for illustrating the implementation of our proposed architecture. The first sub-model is the pre-trained model that was trained with eighty objects, 330,000 total images, from the COCO image dataset. Secondly, the transferred model employing transfer learning technique that train only new image containing three additional classes. There are around 3,000 total images, a thousand images each class. Only several hours are needed for model training. Finally, the ensemble model using bagging technique is used as a decision model to. The experimental results show that the proposed model achieve high performance on test image dataset with 93.33% accuracy.

References

1. Rajeshwari, P., Abhishek, P., Srikanth, P., Vinod, T.: Object detection: an overview. Int. J. Trend Sci. Res. Dev. (IJTSRD) **3**, 1663–1665 (2019)
2. Zhao, Z., Zheng, P., Xu, S., Wu, X.: Object detection with deep learning: a review. IEEE Trans. Neural Netw. Learn. Syst. **30**(11), 3212–3232 (2019)
3. Wu, X., Sahoo, D., Hoi, S.C.H.: Recent Advances in Deep Learning for Object Detection. arXiv preprint, arXiv:1908.03673v1 (2019)
4. Zhang, Y., Sohn, K., Villegas, R., Pan, G., Lee, H.: Improving object detection with deep convolutional networks via bayesian optimization and structured prediction. In: 2015 IEEE Conference on Computer Vision and Pattern Recognition (CVPR), Boston, MA, 7–12 June 2015
5. Erhan, D., Szegedy, C., Toshev, A., Anguelov, D.: Scalable object detection using deep neural networks. In: The IEEE Conference on Computer Vision and Pattern Recognition (CVPR), Columbus, Ohio, 23–28 June 2014

6. Lu, Y., Javidi, T., Lazebnik, S.: Adaptive object detection using adjacency and zoom prediction. In: 2016 IEEE Conference on Computer Vision and Pattern Recognition (CVPR), Las Vegas, NV, USA, June 2016
7. Sun, Q., Pfahringer, B.: Bagging ensemble selection. In: Wang, D., Reynolds, M., (eds.) Advances in Artificial Intelligence. AI 2011. Lecture Notes in Artificial Intelligence, Perth, Australia, 5–8 December 2011, vol. 7106, pp. 251–260. Springer, Heidelberg (2011)
8. Girshick, R., Donahue, J., Darrell, T., Malik, J.: Rich feature hierarchies for accurate object detection and semantic segmentation. In: 2014 IEEE Conference on Computer Vision and Pattern Recognition, Columbus, Ohio, USA, 23–28 June 2014
9. Girshick, R.: Fast R-CNN. In: 2015 IEEE International Conference on Computer Vision (ICCV), Washington, DC, USA, 7–13 December 2015
10. Ren, S., He, K., Girshick, R., Sun, J.: Faster R-CNN: towards real-time object detection with region proposal networks. In: Advances in Neural Information Processing System, Quebec, Canada, December 2015
11. He, K., Gkioxari, G., Dollár, P., Girshick, R.: Mask R-CNN. In: 2017 IEEE International Conference on Computer Vision (ICCV), Venice, Italy, 22–29 October 2017
12. Redmon, J., Divvala, S., Girshick, R., Farhadi, A.: You only look once: unified, real-time object detection. In: 2016 IEEE Conference on Computer Vision and Pattern Recognition (CVPR), New York, USA, 27–30 June 2016
13. O'Shea, K., Nash, R.: An introduction to convolutional neural networks. arXiv preprint, arXiv:1511.08458 (2015)
14. Nwankpa, C., Ijomah, W., Gachagan, A., Marshall, S.: Activation functions: comparison of trends in practice and research for deep learning. arXiv preprint, arXiv:1811.03378 (2018)
15. Lin, T., Maire, M., Belongie, S., et al.: Common objects in context. http://cocodataset.org. Accessed 12 Sept 2019
16. Open Images Dataset V5. https://storage.googleapis.com/openimages/web/index.html. Accessed 21 Nov 2019

Application of Deep Learning to Fairness-Based Power Allocation for 5G NOMA System with Imperfect SIC

Worawit Saetan and Sakchai Thipchaksurat(✉)

Department of Computer Engineering, Faculty of Engineering,
King Mongkut's Institute of Technology Ladkrabang, Bangkok, Thailand
worawit.saetan@gmail.com, sakchai.th@kmitl.ac.th

Abstract. Non-orthogonal multiple access (NOMA) has been a promising technique for 5G communication system, which has higher spectrum efficiency, energy efficiency and fairness than that of orthogonal multiple access (OMA). NOMA serves more than one user by sharing the same time-frequency resource block and uses successive interference cancellation (SIC) to separate multiuser signal. However, the error propagation in the SIC procedure, which is called the imperfect SIC, can cause a severe performance loss. In this paper, we apply the deep learning to power allocation in order to mitigate the impact of imperfect SIC under fairness perspective for downlink NOMA system. Firstly, we formulate an optimization problem aiming to maximize the minimum user rate to provide fairness for all users. Secondly, exhaustive search method is used to solve the formulated problem and thus the optimal power allocation factor is obtained. Lastly, we train the deep neural network to predict the obtained power allocation factor. The simulation results show that our proposed scheme provides the performance close to that provided by exhaustive search. Furthermore, the proposed scheme has much lower complexity than the exhaustive search scheme.

Keywords: Non-orthogonal multiple access (NOMA) · Power allocation · Deep learning · Fairness · Imperfect SIC

1 Introduction

Non-orthogonal multiple access (NOMA) has been a promising multiple access for 5G communication systems. NOMA has drawn more attention from both academia and industry due to its high spectral efficiency, energy efficiency and fairness. The key point of NOMA is to utilize the same resource for serving more than one user via power domain division. Different from conventional orthogonal multiple access (OMA), NOMA supports multiple users through power domain by using superposition coding (SC) at the sender and successive interference cancellation (SIC) at the receiver [1].

P. Meesad and S. Sodsee (Eds.): IC^{2}IT 2020, AISC 1149, pp. 199–207, 2020.
https://doi.org/10.1007/978-3-030-44044-2_20

There have been several researches about NOMA. For instance, In [2], the outage probability was derived with randomly deployed users and fixed power allocation for a downlink NOMA system. In [3], the authors investigated a sum rate maximization problem and provided the solutions to jointly optimize channel and power allocation for NOMA system. In [4], a max-min fairness problem for NOMA system has been addressed. The fairness can be enhanced by adjusting the power allocation factors. In [5], the energy-efficient power allocation of a multiuser NOMA system was investigated. In [6], the joint power and bandwidth allocation have been addressed for maximizing the energy efficiency of the NOMA system.

Most previous researches on NOMA system assumed perfect SIC condition, which represents the users with a stronger channel gain can remove perfectly the interference caused by the users with a weaker channel gain. Nevertheless, in practical systems, error propagation can occur in SIC procedure, which is known as imperfect SIC. Hence, the stronger users remove imperfectly the interference caused by the weaker users. This residual interference can cause a severe performance loss.

Furthermore, a few papers have recently adopted the deep learning to communication field. For instance, In [7], applications of deep learning to wireless communications were introduced in physical layer. In [8], the authors uses deep learning to approximate the weighted MMSE algorithm, which is the optimization algorithm, for maximizing the sum rate.

In order to alleviate the impact of imperfect SIC in practical systems, In this paper, we apply the deep learning to fairness-based power allocation for downlink NOMA system with imperfect SIC. We formulate a power allocation problem with maximizing the minimum user rate to provide fairness for all users. Then, the optimal power allocation factor can be obtained through exhaustive search on the formulated problem. The main idea is to use a deep neuron network (DNN) to learn the exhaustive search scheme.

The rest of this paper is organized as follows. System model and problem formulation are given in Sect. 2. The application of deep learning to fairness-based power allocation is proposed in Sect. 3. The simulation results are presented in Sect. 4. Finally, Sect. 5 concludes the entire paper.

2 System Model and Problem Formulation

2.1 System Model

We consider a downlink NOMA system where one single-antenna base station (BS) serves K single-antenna users over the same time and frequency resource by using power domain multiplexing. The BS is located in the cell center and K users are uniformly distributed within the cell.

By using superposition coding (SC) through power domain division, the transmitted signal at BS can be expressed as

$$x = \sum_{k=1}^{K} \sqrt{p_k} s_k \tag{1}$$

where p_k is the transmit power for k-th user's signal, and s_k is the transmitted signal of k-th user. The received signal at user k can be expressed as

$$y_k = h_k x + n_k, \quad \forall k = 1, 2, \cdots, K \tag{2}$$

where $h_k = g_k \cdot PL(d)$ is the channel coefficient from the BS to user k, g_k is the Rayleigh fading coefficient, and $PL(d)$ is the path loss between the BS and user k at distance d. The last term n_k is the additive white Gaussian noise (AWGN) with variance σ^2.

Besides, we assume that the BS have the perfect channel state information (CSI) and the channel coefficients are sorted as

$$|h_1| \geq |h_2| \geq \cdots \geq |h_K| \tag{3}$$

According to power domain NOMA, SIC procedure is used at the stronger users to alleviate the interference from the weaker users, and thus lead to the better performance. The user k decodes each signal with a weaker channel coefficient, then subtracts it from the received signal. Until finding its own signal s_k, the user k iteratively repeats this SIC procedure in the ascending order of channel coefficients. After finishing SIC procedure, the user k will decode its own signal s_k by treating the remaining signals from the stronger users as noise. If SIC procedure is perfectly done, the user k will cancel interference from the weaker users completely.

However, due to signal-to-interference-plus-noise ratios (SINR), the detection ability of users and modulation and coding schemes (MCSs) in practical systems, the decoded signal of the weaker users at the user k could be imperfect. As a result, the interference cancellation at the user k can also be imperfect. This phenomenon can cause the residual interference at the post-SIC signal of the user k. Hence, The received SINR at the user k can be expressed as

$$\begin{aligned} SINR_k &= \frac{|h_k|^2 p_k}{|h_k|^2 \sum\limits_{i=k+1}^{K} \epsilon_{k,i} p_i + |h_k|^2 \sum\limits_{j=1}^{k-1} p_j + \sigma^2} \\ &= \frac{H_k \alpha_k P_{tot}}{H_k P_{tot} \sum\limits_{i=k+1}^{K} \epsilon_{k,i} \alpha_i + H_k P_{tot} \sum\limits_{j=1}^{k-1} \alpha_j + 1} \end{aligned} \tag{4}$$

where σ^2 represents the noise power, and $H_k = |h_k|^2/\sigma^2$ indicates the channel response normalized by noise (CRNN) of user k, and $p_k = \alpha_k P_{tot}$ represents the transmit power for k-th user's signal, P_{tot} is the total power of BS, α_k indicates the power allocation factor of k-th user's signal, and $\epsilon_{k,i}$ is the uncancelled fraction of the power of a signal of user i that affects user k [9].

Furthermore, the value of $\epsilon_{k,i}$ depends on channel realization obtained by long term measurement. By measuring a huge number of samples, [9] approximated the residual interference as Gaussian distribution. $\epsilon_{k,i}$ can be derived as a variance of the residual interference divided by the received power, which is a feedback to the BS.

Based on the shannon's capacity formula, the achievable rate of user k is written by

$$R_k = Blog_2(1 + SINR_k) \quad (5)$$

where B is the given bandwidth.

2.2 Problem Formulation

In (4), the transmit power can improve the desired signal but can also raise the residual interference. Hence, an inappropriate power allocation can lead to the performance loss of NOMA system. To alleviate the performance loss of the system, we propose a power allocation scheme from perspective of fairness. A common objective of fairness is the maximin fairness (MMF), which aims to provide fairness for all users in the system. The fairness-based power allocation problem can be formulated as the following optimization problem

$$\begin{aligned} max \min_{\alpha_k, 1 \leq k \leq K} &\{R_k\} \\ C1: \alpha_k &\geq 0, \forall k \\ C2: \sum_{k=1}^{K} \alpha_k &\leq 1 \\ C3: R_k &\geq R_k^{min} \end{aligned} \quad (6)$$

where constraint $C1$ ensures that the power for each user is not negative, constraint $C2$ represents the transmit power constraint for BS, and constraint $C3$ indicates that each user rate must be greater than the minimum user rate R_k^{min}.

3 Application of Deep Learning to Fairness-Based Power Allocation

Our proposed scheme applies a deep neural network (DNN) to learn the exhaustive search power allocation (ESPA) scheme, where the input of a DNN is a set of the channel response normalized by noise (CRNN) $\{H_k\}$, the uncancelled fraction of the power of a signal ϵ, and the total power of BS P_{tot}, and the output of a DNN is a set of the power factor $\{\alpha_k\}$. The proposed scheme has 3 hidden layers and each layer contains 20 nodes. 15,000 samples have been generated and divided into a training set containing 12,000 samples and a validation set containing 3,000 samples. The overall process of the proposed scheme is shown in Fig. 1.

In the data generation stage, we use the tuple $(\{\{H_k\}, \epsilon, P_{tot}\}, \{\alpha_k\})$ as the training data, where the total power of BS P_{tot} and the uncancelled fraction of the power of a signal ϵ are in the fixed range, and the channel response normalized by noise (CRNN) H_k is derived from the channel realization with the Rayleigh distribution and the users' locations with the uniform distribution. The optimal power allocation factor α_k is provided through exhaustive search.

In the training stage, we generate multiple tuples $(\{\{H_k\}, \epsilon, P_{tot}\}, \{\alpha_k\})$ and use these tuples to train a DNN. The loss function we use is the mean square error (MSE) between the label $\{\alpha_k\}$ and the output of the DNN. The optimization algorithm we use is the lavenberg-Marquardt algorithm. Further, we use the hyperbolic tangent as activation function for the hidden layers and for the output layer.

In the testing stage, we generate all parameters with the same distribution as the training stage. The tuple $(\{\{H_k\}, \epsilon, P_{tot}\}, \{\alpha_k\})$ is used as the testing data. We send each tuple into the trained DNN and then collect the predicted power allocation factor. Finally, We evaluate the performance metric of the trained DNN and compare it with that of the exhaustive search power allocation (ESPA).

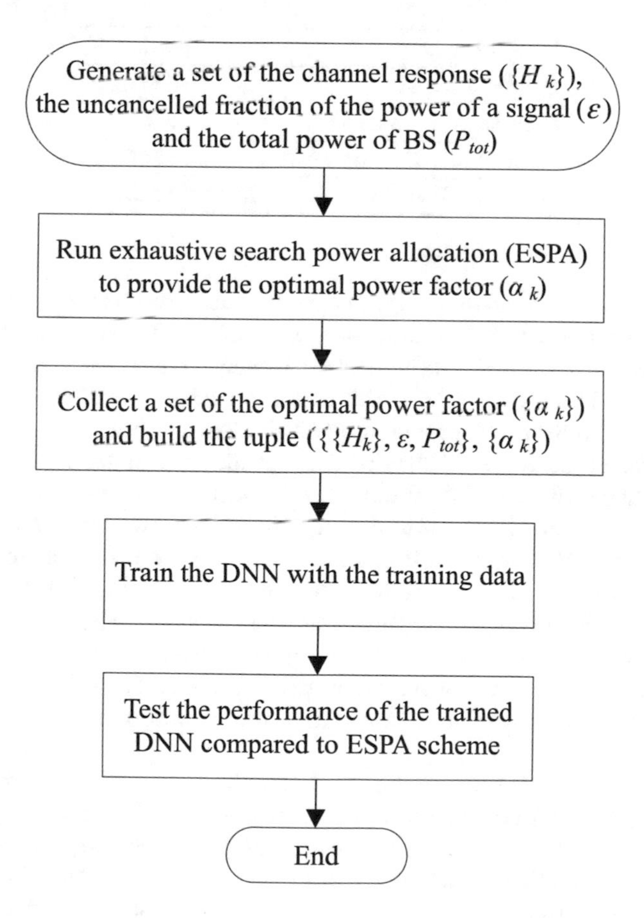

Fig. 1. The overall process of our proposed scheme

4 Simulation Results

In this section, we evaluate the performance of our proposed scheme for NOMA system through simulation. We compare our proposed scheme with exhaustive search power allocation (ESPA) scheme, which is to find the solution of the optimization problem in (6). ESPA scheme provides the best performance but is not appropriate for practical system due to its high complexity. We consider one base station located in the cell center and the K users are distributed randomly within 300 m. We define the minimum distance between the base station and the users as 40 m. For simplicity, all $\epsilon_{k,i}$ are equal to ϵ [9] and all R_k^{min} are equal to R^{min}. Table 1 depicted the simulation parameters.

Table 1. Simulation parameters

Parameter name	Value
Bandwidth (B)	5 MHz
Path Loss Exponent (v)	2.3
Noise Power Density (N_0)	174 dBm
Minimum User Data Rate (R^{min})	1 Mbps

For 2-users NOMA system, Fig. 2 compares our proposed scheme with ESPA scheme. We evaluate the minimum user rate of the NOMA system versus the power of BS. The proposed scheme can closely attain the optimal minimum user rate. For the case $\epsilon = 0.00$, the proposed scheme achieves 98.77% of the optimal minimum user rate provided by ESPA scheme. For the case $\epsilon = 0.01$, the proposed scheme achieves 97.90% of the optimal minimum user rate provided by ESPA scheme. In addition, due to the impact of the uncancelled fraction of the power of a signal, the minimum user rate of both two schemes can decrease.

For 3-users NOMA system, Fig. 3 compares our proposed scheme with ESPA scheme. We evaluate the minimum user rate of the NOMA system versus the power of BS. The proposed scheme can closely attain the optimal minimum user rate. For the case $\epsilon = 0.00$, the proposed scheme achieves 96.99% of the optimal minimum user rate provided by ESPA scheme. For the case $\epsilon = 0.01$, the proposed scheme achieves 96.49% of the optimal minimum user rate provided by ESPA scheme.

For 2-users NOMA system, Fig. 4 shows the average CPU time versus the power of BS. We compare our proposed scheme with ESPA scheme. In this figure, we set $\epsilon = 0.01$. Obviously, the proposed scheme takes the average CPU time much less than ESPA scheme. For example, when the power of BS is 90 mW, the average CPU time of the proposed scheme is 0.0564 ms but that of ESPA scheme is 6.9310 ms, which is approximately 122.89 times. Besides, as the power of BS increases, the average CPU time of the proposed scheme remains constant while that of ESPA scheme grows exponentially.

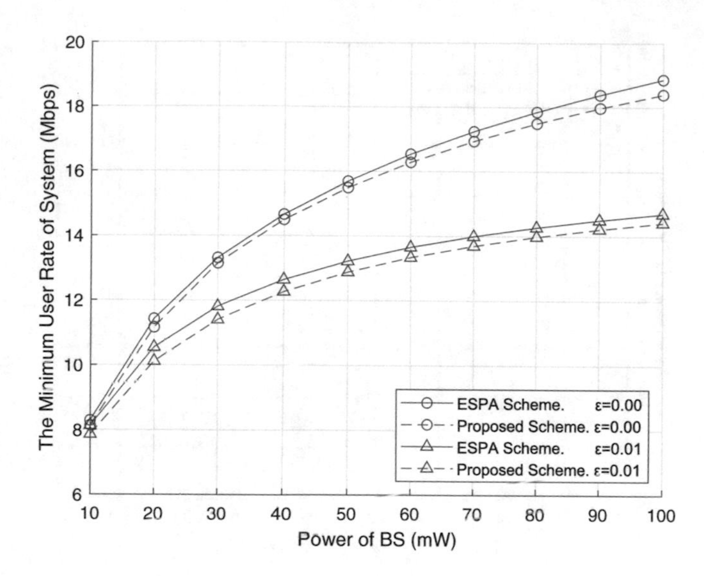

Fig. 2. The minimum user rate of the proposed scheme and ESPA scheme for 2-users NOMA system.

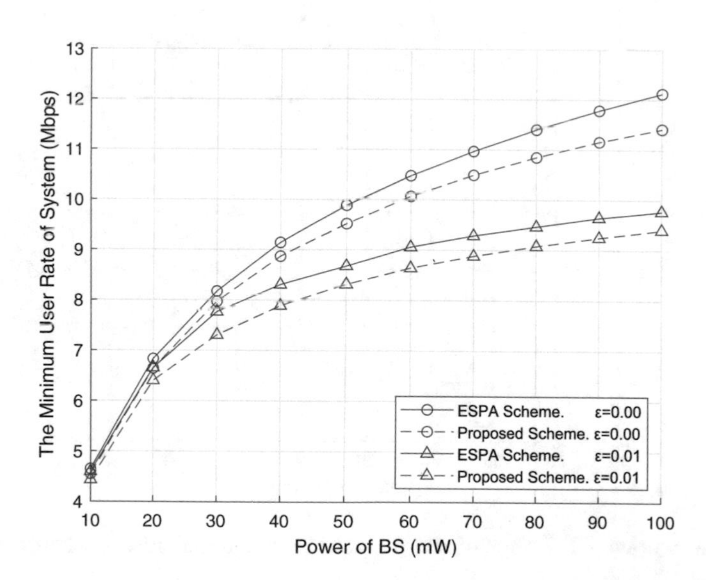

Fig. 3. The minimum user rate of the proposed scheme and ESPA scheme for 3-users NOMA system.

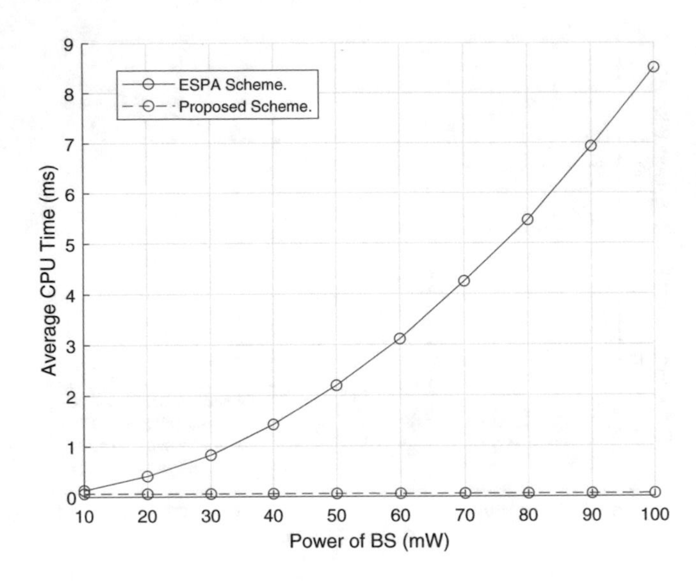

Fig. 4. The average CPU time of the proposed scheme and ESPA scheme for 2-users NOMA system.

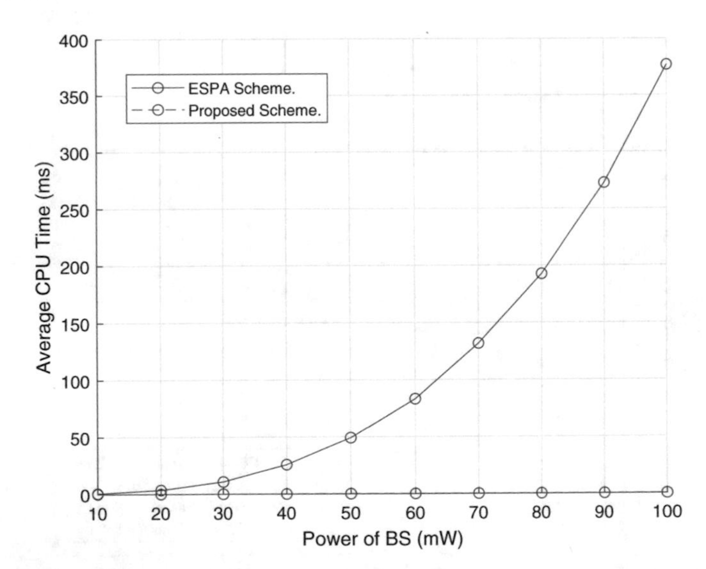

Fig. 5. The average CPU time of the proposed scheme and ESPA scheme for 3-users NOMA system.

For 3-users NOMA system, Fig. 5 shows the average CPU time versus the power of BS. We compare our proposed scheme with ESPA scheme. In this figure, we set $\epsilon = 0.01$. Obviously, the proposed scheme takes the average CPU time much less than ESPA scheme. For example, when the power of BS is 90 mW, the average CPU time of the proposed scheme is 0.1209 ms but that of ESPA scheme is 272.5052 ms, which is approximately 2253.97 times. Besides, as the power of BS increases, the average CPU time of the proposed scheme remains constant while that of ESPA scheme grows exponentially.

5 Conclusion

In this paper, we applied the deep learning to power allocation and proposed fairness-based power allocation scheme for downlink NOMA system in the presence of imperfect SIC. The optimization problem was formulated to maximize the minimum user rate of the system. We found the solution of the formulated problem by using the exhaustive search and then designed the deep neural network to learn the exhaustive search power allocation (ESPA). The simulation results have shown that the proposed scheme with much lower complexity provides the minimum user rate close to that provided by ESPA scheme.

References

1. Islam, S.M.R., Avazov, N., Dobre, O.A., Kwak, K.: Power-domain non-orthogonal multiple access (NOMA) in 5G systems: potentials and challenges. IEEE Commun. Surv. Tutor. **19**(2), 721–742 (2017)
2. Ding, Z., Yang, Z., Fan, P., Poor, H.V.: On the performance of non-orthogonal multiple access in 5G systems with randomly deployed users. IEEE Signal Process. Lett. **21**(12), 1501–1505 (2014)
3. Lei, L., Yuan, D., Ho, C.K., Sun, S.: Power and channel allocation for non-orthogonal multiple access in 5G systems: tractability and computation. IEEE Trans. Wirel. Commun. **15**(12), 8580–8594 (2016)
4. Timotheou, S., Krikidis, I.: Fairness for non-orthogonal multiple access in 5G systems. IEEE Signal Process. Lett. **22**(10), 1647–1651 (2015)
5. Zhang, Y., Wang, H., Zheng, T., Yang, Q.: Energy-efficient transmission design in non-orthogonal multiple access. IEEE Trans. Veh. Technol. **66**(3), 2852–2857 (2017)
6. Wang, J., Xu, H., Fan, L., Zhu, B., Zhou, A.: Energy-efficient joint power and bandwidth allocation for NOMA systems. IEEE Commun. Lett. **22**(4), 780–783 (2018)
7. O'Shea, T., Hoydis, J.: An introduction to deep learning for the physical layer. IEEE Trans. Cogn. Commun. Netw. **3**(4), 563–575 (2017)
8. Sun, H., Chen, X., Shi, Q., Hong, M., Fu, X., Sidiropoulos, N.D.: Learning to optimize: training deep neural networks for wireless resource management. In: 2017 IEEE 18th International Workshop on Signal Processing Advances in Wireless Communications (SPAWC), pp. 1–6, July 2017
9. Agrawal, A., Andrews, J.G., Cioffi, J.M., Meng, T.: Iterative power control for imperfect successive interference cancellation. IEEE Trans. Wirel. Commun. **4**(3), 878–884 (2005)

Author Index

P. Meesad and S. Sodsee (Eds.): IC^{2}IT 2020, AISC 1149, pp. 209–210, 2020.
https://doi.org/10.1007/978-3-030-44044-2